The Transit of Venus

By the same author

Endeavour:
The Story of Captain Cook's First Great Epic Voyage

Newton's Apple:
Isaac Newton and the English Scientific Renaissance

North Meols and Southport

Liverpool: A People's History

Bristol: A People's History

Resolution:
Captain Cook's Second Voyage of Discovery

The Transit of Venus

The Brief, Brilliant Life of Jeremiah Horrocks, Father of British Astronomy

PETER AUGHTON

Weidenfeld & Nicolson

LONDON
A Windrush Press Book

First published in Great Britain in 2004
by Weidenfeld & Nicolson
in association with The Windrush Press
Second impression May 2004

A CIP catalogue record for this book
is available from the British Library.

ISBN 0 297 84721 X

Printed in Great Britain by Butler & Tanner Ltd,
Frome and London

Weidenfeld & Nicolson

The Orion Publishing Group Ltd
Orion House
5 Upper Saint Martin's Lane
London, WC2H 9EA

The Windrush Press
Windrush House
Adlestrop
Moreton in Marsh
Glos GL56 0YN

To Jackie

CONTENTS

FIGURES

ILLUSTRATIONS

PREFACE

When I first conceived the idea of writing a biography of Jeremiah Horrocks, I had very little idea about how the book would turn out. I did not even know whether enough data survived for a biography, or where I could find that data if it existed. Horrocks has long been acknowledged as a significant but obscure figure in the history of science and as a slightly better known figure in the more specialized field of astronomy. He warrants a mention in the *Dictionary of Scientific Biography*, but he remains a shadowy figure because very little has been published about him and there are no readily available sources on his life and works.

About twenty years ago, I was very fortunate to obtain a copy of the Reverend A. B. Whatton's *Memoir of the Revd Jeremiah Horrox*, published in 1859. The title itself is misleading. It uses the Latinized spelling of his name and it makes the unproven assumption that Jeremiah Horrocks was ordained. Even so, the book is a rare and valuable source. I have made full use of Whatton's work and have quoted from his translation of Horrocks's *Venus in Sole Visa*, which includes Jeremiah Horrocks's poetry, almost without modification.

When I came to search further, I discovered that Whatton's book was the only biography of Jeremiah Horrocks ever to be published. Throughout the remainder of the nineteenth century and the whole of the twentieth, no further biographies appeared. A number of articles were published in the astronomical journals, however, and I am indebted to the library of the University of the West of England for obtaining copies of these articles to assist me with my research. I knew that more than three centuries ago John Wallis at the Royal Society was engaged in a similar task of finding and publishing the 'remains of Horrocks' and that the *Opera Posthuma of Jeremiah Horrox*, edited by John Wallis, was published in 1672. This was a main source of information, and I therefore had to find a copy and spend time studying it. My only option was to make a direct approach to the Royal Society. The response was very encouraging and I have to acknowledge the assistance of Rupert Baker, the Royal Society librarian, who not only produced a copy of the book for me but was also able to arrange microfilming so that I could study the original text at my leisure.

Other sources suggested that there might be some original documents in the archives of the Royal Greenwich Observatory. Again I struck gold. Thanks to Adam Perkins, I was able to obtain microfilm copies of Horrocks's original writings, in particular his 'Astronomical Exercises' and his 'Philosophical Exercises'. These notebooks were written in English and I have quoted freely from them, using Horrocks's own words. Discerning readers will be able to spot that the English in *Venus in Sole Visa,* which includes Horrocks's poetry, is more modern than the English quoted from Jeremiah Horrocks's notebooks. The reason for this is that the former was translated by Whatton from the Latin into Victorian English, but the latter are quoted in the seventeenth-century English of Horrocks's own times.

Another valuable and original source of information was the Lancashire Records Office. A few parish register entries were valuable in putting together the family trees, but the registers from Horrocks's time are very incomplete. Fortunately, many members of both the Horrocks and Aspinwall families left wills, and this has enabled me to piece together both family trees with confidence. These wills also provide information on the early watchmaking industry and I want to mention Dennis Moore at the Prescot Museum of Clockmaking, who kindly gave me copies of his own genealogical research, compiled from watchmaking sources, which was in good agreement with my own work.

Another original source, which became available at exactly the right time, was the recently published correspondence of John Flamsteed, the first Astronomer Royal. There are many references to Horrocks in these fascinating volumes, and I was fortunate to be given some assistance by one of the co-authors, Frances Wilmoth of Cambridge University Library. It was very heartening to find that both the Royal Society and the Royal Greenwich Observatory recognized that a biography of Jeremiah Horrocks was long overdue.

I am also indebted to Dr H. C. Carron and Mrs Janet Morris of Emmanuel College, Cambridge. It was through their recommendations that I was able to discover the unexpected connections between Emmanuel College and the Massachusetts Bay Company. I must also mention Professor Rodney Sampson of Bristol University, who has helped me with selected Latin to English translations from the *Opera Posthuma*.

Thus, a very high proportion of this book has been compiled from original seventeenth-century sources and much of it will be new to both astronomers and historians. Fortunately, the astronomy is so early that it is easily comprehensible to the layman and much of it can be repeated by the amateur today. Finally, all the work would have been to little avail without the efforts of Victoria Huxley and colleagues at Weidenfeld and Nicolson, who agreed to publish this book.

Peter Aughton
Bristol 2002

CHAPTER ONE

The Royal Society

There was great consternation at the Royal Society. The Polish astronomer Johannes Hevelius had published an astronomical paper from his observatory at Danzig. It was a fascinating paper and it described an event from over twenty years ago, namely the one and only transit ever to be observed of the planet Venus across the Sun. The paper was quite unlike any scientific paper ever seen before. It was elegantly written and full of youthful enthusiasm. It contained humour and even poetry. The author was a brilliant astronomer who had used his measurements of the transit not only to correct the orbit of Venus but also to estimate a value for the distance from the Earth to the Sun. It was an amazing paper by any standards, but the problem which faced the Royal Society was that it had been written by an Englishman. How had it come about that such an elegant and valuable publication had first appeared in a foreign country?

There was much discussion. Christopher Wren, the professor of astronomy at Oxford University, was consulted. He confirmed the great merit of the treatise and the observation. There were a few problems and conclusions with which the learned society begged to differ. For example, the author put the Sun much further away than any other astronomer: a distance of over sixty million miles from the Earth to the Sun had been calculated. The writer also thought that the planets Jupiter and Saturn were so large that they were much greater than the Earth. But his reasoning had a compelling logic. Who was the author of this marvellous paper on astronomy? He was a Mr Horrocks who, according to the information in the publication, had made the observation of Venus from a place eighteen miles to the north of Liverpool. He was long deceased, but surely there must be somebody in England who knew something about him?

When enquiries were made, there were at least two people who knew a little about Jeremiah Horrocks. One was John Wallis, an active and respected member of the Royal Society. Back in the 1650s, Wallis had been a member of the pioneering Oxford Philosophical Society. He was a mathematician who had become a founder member and a leading light at the Royal Society. Another man who had known Horrocks was the Revd John Worthington, who held the living of Barking in Essex and also the living of Needham in Suffolk. He was a regular preacher at St Benet Fink's in London and he was well known in academic circles as a former vice chancellor of the University of Cambridge. Both these men had known Jeremiah Horrocks in their student days at Cambridge, before the war broke out. When John Wallis first saw the paper written by his old college friend, he naturally took a great interest in it. His comments echoed the view of the Royal Society:

> I cannot help being displeased that this valuable observation, purchasable by no money, elegantly described, and prepared for the press, should have laid hid for two-and-twenty years, and that no one should have been found to take charge of so fair an offspring at its father's death, to bring to light a treatise of such importance to astronomy, and to preserve a work for our country's credit and for the advantage of mankind.[1]

Yes, John Worthington and John Wallis both remembered their friend Jeremiah. All three had been colleagues together at Emmanuel College. They remembered the carefree days of their youth, the long evenings when the day's chores were over, when they put the world to rights and discussed what they were going to do with their lives and their futures. Those halcyon days before the war when there was peace and prosperity, when political conflict was no more than a debate over a glass of ale in the taverns and colleges. They remembered the freedom of university life when the world was a younger and better place. From their college days they remembered the great preacher John Cotton who moved from Boston, Lincolnshire to Boston, Massachusetts. They remembered the academic John Harvard whose legacy had founded the first university in the New

World. They remembered others from their college who had left England on those small, frail vessels to found the thriving colony of Massachusetts on the other side of the Atlantic Ocean. They had all wanted to change the world for the better. Yes. Jeremy Horrocks. A boy full of the joy of life, a youth as much at home with his poetry as with his beloved astronomy. He alone at Cambridge had had horizons even further away than the New World. Even then, his ambition was to study the heavens, to measure the stars, to follow the wanderings of the planets in the skies. In those days, he was the only person at Cambridge to believe in the revolutionary theory of Copernicus which put the Sun at the centre of the universe, with the Earth orbiting around it. He wanted to discover more about the universe. He wanted to know how and when the world had been created. He wanted to understand the forces which drove the planets around the Sun. He wanted to know how far it was to the Sun and the distance to the stars. He wanted to measure the Moon and the stars at a time when the telescope was a new invention. He wanted to expand his chosen subject of astronomy as far as it was possible to go in those nostalgic times, before the outbreak of the Civil War upset the peace of old England. His work was a whole generation ahead of the establishment of the Royal Society.

John Worthington actually knew something about the works of Jeremiah Horrocks before the Royal Society became involved. Worthington had been instrumental in getting the paper on the transit of Venus published. For several years, he had been trying to obtain information about his old college friend. His quest began in the 1650s and his enquiries led him to Broughton, near Manchester, where he knew that another astronomer called William Crabtree had been a great friend and correspondent of Jeremiah Horrocks. Crabtree, too, was gone, having died soon after the battle of Marston Moor, but Worthington discovered some astronomical papers amongst Crabtree's estate. They had been written by Jeremiah Horrocks and he was able to purchase them. One of these papers was Horrocks's account of the transit of Venus in 1639. In fact, it seems there were two drafts of the paper. A few years after making his find, on 28 April 1659, Worthington wrote a letter to Samuel Hartlib, a man he knew to be well connected in scientific publishing. He enclosed both copies of Horrocks's papers, explaining how and where he had come by the manuscripts:

> I have, as you desire, sent you Mr. Horrox, his discourse called 'Venus in sole visa'. Here are two copies of it, but neither writ to the end. I lent them some years since to a friend who promised out of both to make out one, and then to print it; but other business it seems would not permit him to go through with the work. In some other loose papers I perceive that the author began his tract again and again (so curious was he about it), but these seem to be his last, written with his own hand. He lived at Toxteth Park near Liverpool, in Lancashire, was some time of Emmanuel College, Cambridge, admitted the same year I was. These papers of his, with many others of astronomical observations, I found in the study of one Mr. Crabtree (a Lancashire man, and his great correspondent in these studies), and I bought them after his death. By sending to some friend about Liverpool or Toxteth, it may be known whether any of Mr. Horrox's kindred have any of his papers.
>
> Yours, &c
>
> J. Worthington [2]

Worthington does not seem to have visited Jeremiah's birthplace at Toxteth. Many of Jeremiah's family were still living at this time. Their business of watchmaking was prospering and they had expanded their trade to London. The area around Liverpool had suffered badly from looting during the Civil War and the evidence implies that the looting had destroyed all that was left of Horrocks's work in Toxteth. He had a brother, Jonas, who, it seems, had fled to Ireland to escape from the war, taking some of Jeremiah's papers with him. The papers had been lost and never seen again. None of his work seemed to survive from Jeremiah's home at Toxteth.

But Samuel Hartlib was very interested in the paper on Venus which John Worthington showed him. He showed it in turn to the prominent Dutch astronomer, Christian Huyghens, a respected scientist who was not slow to appreciate the value of the work. The interest of the European scientific world became aroused. Huyghens was a great friend of the Polish astronomer Hevelius and he knew that the latter was preparing to publish a paper on a transit of Mercury across the Sun, an event which he had recently observed at Danzig. Huyghens knew that Hevelius would jump at the opportunity to include an account of the transit of Venus with that of Mercury.

In the meantime, Samuel Hartlib took a long time to return the copies of 'Venus in Sole Visa' to John Worthington – so much so that Worthington was moved to write and ask for their return: he was concerned that they might be the only surviving copies of his friend's work. He need not have feared. When Hevelius received the manuscript, he expressed great satisfaction that the two tracts, his own and that of Horrocks, were to be made into one volume. His wording was fulsome and elegant:

> How greatly does my Mercury exult in the pious prospect that he may shortly fold within his arms Horrox's long-looked for, and beloved Venus. He renders you unfeigned thanks that by your permission this much-desired union is about to be celebrated, and that the writer is able with your concurrence to introduce them both together to the public.[3]

Hevelius was as good as his word, publishing both treatises for the whole scientific world to read. He ensured that at least this part of Horrocks's work would be known to posterity. It was May 1662 when Hevelius sent the newly published treatise to Huyghens. His accompanying letter was coloured by the fact that he was in mourning, having recently lost his wife, whom he refers to as 'my own beloved Venus'. He apologized for the delay that his bereavement had incurred:

> You have doubtless heard, much honored friend, of the severe domestic calamity by which I was prevented from more quickly fulfilling my promise; and I am sure you will not only readily excuse me, but sympathize with me in this trial, when you understand how grievous an affliction has befallen me. I have sent you by Dr. Peltrius my Mercury produced amidst great mental anxiety, together with Horrox's Venus, happily risen for the public good, whilst alas! my own beautiful Venus has set to my infinite sorrow! I pray you to consider them carefully, until I am able to send you something better. The learned world is particularly

> indebted to you for bringing Horrox's Venus to light, thus having cheerfully bestowed a gift so excellent and acceptable as to demand the thanks of the latest posterity. When you have read the book, I beg you will give me your opinion of its merits, which I shall esteem a great kindness, and in turn you will always find me desirous of serving you.[4]

The reply from Christian Huyghens to Hevelius was written on 25 July 1662. He sent his thanks to Hevelius and his sympathies over the bereavement. Hevelius had managed to publish the work in spite of the grievous loss of his wife and Huyghens greatly appreciated his efforts:

> Your most acceptable letter, and shortly afterwards the volume of the new observations reached me safely, and although I ought to have thanked you before for the valuable gift, I have been so hindered that I could not until now discharge this duty. The illustrious Bullialdus informed me of the great affliction you have sustained by the death of your dearest wife, on which account I feared that this little work, which was then in hand, would be delayed. But you have acted rightly in not suffering your private loss to become a public misfortune: for I cannot say how highly astronomy is indebted to you for so accurate a description of your beautiful observation. Posterity cannot adequately repay you with its thanks. Touching the posthumous work of Horrox now brought to light, it is more satisfactory that it should have been undertaken by you, than by me; especially as you have prepared an excellent and elegant edition, and increased its value by a commentary.[5]

It was thus John Worthington who set up the chain of events which led to Jeremiah Horrocks coming to the notice of the Royal Society. The response of the society, when they discovered the merits of their deceased countryman, can only be described as moving and magnificent. They knew that there must be other works by this remarkable astronomer and they made a unique decision to publish posthumously the works of a scientist who had been dead for a whole generation. There was something romantic and emotional about a young genius cut off in his prime, but their decision involved more than pure sentiment. The Royal Society realized that many years before its foundation there had been in England a man who had fully appreciated their aims and *raison d'être.* He was a man who had sprung from nowhere but who understood instinctively the differences between the scientific method which the society existed to promote and the charlatan world of alchemy, astrology, magic and witchcraft in which the great majority of people still believed.

The search continued for any surviving works by this young astronomer. The plan was to collect them, prepare them for the press and publish them under the title of 'Posthumous Works' – the *Opera Posthuma* of Jeremiah Horrocks. In John Wallis the Royal Society had the ideal man to edit this publication. He was a man in his mid-forties, not only a founder member of the society, but most importantly he had a personal interest. He was, like John Worthington, a friend of Jeremiah Horrocks from college days.

There were other parties who were interested in the quest for the remains of the works of Horrocks. One of these was a young man called John Flamsteed. Whereas Worthington and Wallis were contemporaries of Horrocks, Flamsteed was not born until five years after his death. He was only seventeen when *Venus in Sole Visa* was published but by the time the *Opera Posthuma* was being prepared for the press he was in his twenties and had already made his reputation as an excellent astronomer. He was destined to become England's first Astronomer Royal. Flamsteed had read *Venus in Sole Visa* and understood it better than anybody else alive, even John Wallis. He deserves more than a passing mention as one of the few who rescued the works of Jeremiah Horrocks from oblivion.

The Royal Society was based in London, but saw itself as a national society and had many country members. One of these was Richard Towneley of Towneley Hall, near Burnley in Lancashire. The Towneleys were wealthy landowners and dedicated scientists, but they were Catholic recusants at a time when the Puritan movement was in the ascendancy and for this reason they tended to keep a low profile. They were among the first in England to make systematic recordings of rainfall and meteorological phenomena and as a result of these experiments Richard Towneley has a place in the history of science. It was he who discovered a simple relationship between the pressure and volume of a fixed mass of gas at constant temperature. Robert Boyle called this 'Towneley's Law' and he made much of it in his own experiments – so much so that it became known as 'Boyle's Law'. This was not a deliberate act of plagiarism on the part of Boyle; his fellow members at the Royal Society insisted on naming the law after him.

Flamsteed knew that some of Horrocks's manuscripts had found their way to Towneley Hall. Once again, the war had much to answer for. Charles Towneley, father of Richard, had been killed at the battle of Marston Moor in 1644, in the company of many other royalists. At this time, some of Horrocks's astronomical papers were in the possession of Charles's brother, Christopher Towneley. Flamsteed decided to make the journey to Lancashire. After several days' travelling, he arrived at his destination only to find that Christopher Towneley was in London. But Flamsteed's journey was not in vain:

> I made a journey into Lancashire, and called at Townley, to visit Mr. Christopher Townley, who happened to be then in London. But one of his domestics kindly received me, and shewed me his instruments, and how his micrometer was fitted to his tubes; and from this time forward we often conferred by letters. I procured Mr. Gascoigne's and Mr. Crabtree's papers from him, and Mr. Horrox's theory of the moon, to which he had begun to fit some numbers; but perfected none that I remember. About this time Mr. Horrox's remains and observations, having been collected by Dr. Wallis, were in the press. I found his theory [of the lunar motion] (of which a correct copy had fallen into my hands) agree much better with my observations than any other. Hereupon I fitted numbers to it, which with an explanation of it were printed with his works. Mr. Collins advised me to print my discourse concerning the equation of natural days with them: which I consented to do; and sent it up to him for that purpose translated into Latin.[6]

Flamsteed was delighted to discover one of the objects of his quest: more of Jeremiah Horrocks's papers. The account of his theory of the Moon's motion was a very significant find in that years later, when Flamsteed became Astronomer Royal, he would have to find a way to predict the Moon's motion in order to solve the problem of finding the longitude at sea. John Flamsteed also discovered the data for an eclipse of the Sun which took place on 22 May 1639, observed by William Crabtree. There were observations of an eclipse of the Moon on 10 December 1638 by William Gascoigne, the inventor of the micrometer which was fitted to Towneley's telescope. The lunar eclipse had also been observed by Jeremiah Horrocks and he had tried to use the data to calculate the longitude of a point on the American continent by comparing his timings with the observations of Thomas Horrocks, his American cousin. There were, of course, many observations of solar and lunar eclipses which had been recorded elsewhere, but Flamsteed knew the value of accurate observations from the past. They could never be repeated and they gave information and positions which could be used by future astronomers. He was able to purchase the papers and they eventually found their way into the collection at the Royal Greenwich Observatory. He made himself known to John Wallis, who was responsible for the publication of Horrocks's *Opera Posthuma.* He wanted his new findings to be added to the publication.

Flamsteed was told that more of the papers had survived but that they were no longer in the north of England. They had been transferred to the safe keeping of Jonas Moore at the Tower of London. Flamsteed was working from his home in Derby, but he was a regular visitor to London and he had contacts at the Royal Society. He was already indebted to Jonas Moore and did not think he could persuade him to part with the Horrocks documents, even for publication by the Royal Society. He wrote to John Collins of the Royal Society, asking for his support. Collins lived at William Austin's house against the Adam and Eve in Petty France, Westminster:

> Sir
> with these I send you my promisd Solar tables by my scholefellow Mr Sargeant: tho I have not added the Suns diameters yet I intend before these may be printed to send them to you: to be annexed to them or included with the rest: Sir I understood by Mr Townlys Unkle [ie Christopher, uncle of Richard] that Mr Jonas Moore has severall of Mr Gascoignes and Mr Crabtrees papers or letters which having one part of them in my hands I would gladly see

> because they often make reference to some others that I have not and which Mr Christopher Townly thinkes are in Mr Moores hands, but I do not know how to move Mr Moore about them I am an egregious debtor to him already and such manuscripts I know to be a treasure of that nature as is not to be trusted to every one. And I can not tell how with civility to demand them of him, but you haveing given mee notice of these and some information of his I would entreat you when you have a convenient opportunity that you would please to speake to him about them what your owne discretion may dictate to you so that they may come safe to my hands: I desire to know whether some partes or the summe of them have not beene excerpt by Dr Wallis into his collection of Mr Horroxes papers, for I find that there was a constant intercourse of mathematicall dissertations betwixt Mr Crabtree and Gascoigne and that it reacht to Mr Horrox too, tho Gascoigne and Crabtree became acquainted but some little while before Mr Horrox death: I esteeme Mr Gascoigne by his papers to have beene as ingenuous a person as the world has bred or knowne yet better versed in Mechanic inventions and happier in them then in his Astronomy . . . [7]

The picture was beginning to come together. Much had been lost, but what remained was well worthy of publication, even though it had been written a generation before. The truth was quite incredible. Back in the late 1630s, before the outbreak of the Civil War, a young genius called Jeremiah Horrocks had tackled and solved the most difficult astronomical problems of the age. He was a rare breed of astronomer, not only skilled at observational astronomy, but also in theoretical and philosophical work. He knew the positions and the motions of the planets more accurately than any other astronomer of his time. He was the first man to see the image of Venus on the face of the Sun. He was the first man to formulate a valid theory for the wanderings of the Moon. He was the first man to appreciate the true scale of the solar system. Yet he was not an elderly, grey-bearded sage. He was a young man full of youth and vitality and he had barely come of age.

CHAPTER TWO

Toxteth

Strict fines were imposed on those sects who did not support the established Church of England. Bishop Bridgeman of Chester sent out regular visitations to weed out the recusants who did not attend church. The diocese of Chester included much of the neighbouring county of Lancashire, a hotbed of Catholic recusants, many of whom were the landowners and gentry. At this time, long before the Industrial Revolution, Lancashire was a somewhat isolated and very rural county. It had never seen a bale of cotton. Secret masses and clandestine marriages were common in many of the manor houses, and there were priest's holes in which to harbour Catholic priests and conceal them from the Church authorities. Catholicism had always been present in the county – after the Reformation, many families had never abandoned the old faith – but in more recent times there had been a movement away from Catholicism. The Puritans were making great progress, particularly in the area around Manchester and Bolton and to a lesser extent in the area around Liverpool. Their mission was to 'purify' the faith and to keep it as simple as possible. One example was that they did not hold with the Catholic practice of worshipping holy relics and graven images. Bishop Bridgeman, who had enough problems to deal with from the Catholics, was prepared to condone the Puritans as a lesser evil. He would have been happy to turn a blind eye to their diversions from the established religion, but the matter had come to the ears of the Archbishop of Canterbury, who would have none of it. There were two nonconformist ministers in the Bolton area: John Angier and his neighbour, the Revd Alexander Horrocks of Deane, who had decided to serve communion with his parishioners standing up instead of meekly kneeling. Bishop Bridgeman wrote to John Angier with great reservations but giving him his marching orders. Alexander Horrocks is mentioned in the same letter:

> I have a good will to indulge you but cannot, for my Lord's Grace of Canterbury hath rebuked me for permitting two nonconforming ministers, the one within a mile on one hand, Mr Horrocks of Deane, on the other yourself, and I am likely to come into disfavour on this behalf. As for Mr. Horrocks my hands are bound, I cannot meddle with him, but as for you, Mr Angier, you are a young man and may doubtless get another place; and if you were anywhere a little further distance I could better look away from you, for I do study to do you a kindness, but cannot as long as you are thus near me.[1]

It seems that Alexander Horrocks had good family connections and the bishop was unable to dismiss him so easily because of an unknown promise made to his wife. Alexander remained a notorious nonconformist at Westhoughton and Deane for another thirty years. During the Civil War, he was well known to the Cavaliers, who described him as 'That old rogue who preaches in his grey cloak'.

On 17 January 1615, the Revd Alexander Horrocks performed a wedding ceremony at the church of Deane between his cousin James Horrocks and a spinster called Mary Aspinwall from Toxteth Park, near Liverpool. The bondsman was Peter Ambrose, brother-in-law of the bride. It should perhaps be mentioned that Horrocks family weddings did not always go to plan. A few years previously, when Margaret Horrocks arrived at Middleton church to marry a man by the name of Seddon, the groom turned up at the church but could not face the consequences. The record states that 'hee fled [the church] when he should have sealed the bond'. This time, however, the ceremony was successfully performed without either party fleeing the church.

The parish registers show that the most numerous branches of the Horrocks family were in the Bolton area of Lancashire, where James Horrocks and Mary Aspinwall were married. The name originated in the central region of the county under the spelling 'Harrocks', and the family claimed descent from the Harrocks of Harrocks Hall, an edifice which stands above the Douglas Valley between Parbold and Eccleston. The name is thought to be a derivation of the place name 'Hoary Oak'.

The Aspinwall family originated from a neighbouring part of Lancashire, between Ormskirk and Scarisbrick, where they had been landowners since the Middle Ages. The original spelling was 'Aspen Well' and the name was therefore also derived from a species of tree. Both families were well established in the seventeenth century and the occasional Horrocks or Aspinwall is mentioned as a 'gentleman' or 'esquire', but for the most part they had fallen below the ranks of the landed gentry. Both families were well educated, however, and both have entries among the alumni of both Oxford and Cambridge. Younger sons on the Horrocks side favoured Cambridge and they appear in the college registers of Emmanuel, Christ Church and St John's. The Aspinwalls favoured Oxford, where Brasenose was the main college of their choice.

James and Mary Aspinwall did not settle in the Bolton area where they were married, but moved instead to live in Aspinwall country at Toxteth Park, though they remained in close touch with their Bolton relatives. James was a watchmaker, and it seems probable that he originally moved to Toxteth Park as a single man to start his watchmaker's apprenticeship. This would have been in about the year 1610 and it was there that he met his future wife. Circumstances suggest that he was apprenticed to the watchmaker Thomas Aspinwall, the father of the bride, and that he therefore married his master's daughter. Three years later, the marriage was blessed with a son, and a second son was born three years after the first. The parents had a penchant for the names of Old Testament prophets of doom. Their first son, Jeremiah, was born in 1618. The second, christened Jonas or Jonah, was born in 1621.

The registers of St Nicholas Chapel in Liverpool do not record the baptism of Jeremiah Horrocks, nor do the registers of the neighbouring churches of Childwall and Walton on the Hill, but all three registers are incomplete for the critical years in question. There is a possibility that the brothers were baptized at the new chapel in Toxteth, a Puritan meeting house which was first established in 1618, the year Jeremiah was born, but for which no contemporary baptismal records survive. The Bolton registers record a Jonas Horrocks, son of James, who was baptized in February 1622 and who could therefore be Jeremiah's younger brother. This evidence suggests that the parents did not move to Toxteth until 1622 or later and that Jeremiah was born in Bolton. The lack of records leaves us with no proof that Jeremiah Horrocks was born in either place, but his parents were certainly established at Toxteth soon after their marriage and the tradition that Jeremiah lived and spent his childhood there is well founded. Toxteth, in the twenty-first century, is remembered as the scene of riots and inner-city deprivation. It was not always thus. In the seventeenth century, Toxteth covered a much greater area and was exclusively rural. In 1792, William Roscoe described the Dingle, a leafy valley which was part of Toxteth:

> Stranger, that with careless feet,
> Wandered near this green retreat,
> Where, through gently bending slopes,
> Soft the distant prospect opes;
>
> Where the fern in fringed pride
> Decks the lonely Valley's side;
> Where the linnet chirps his song,
> Flitting as thou tread'st along.
>
> William Roscoe[2]

Lawrence Hall, quoting an extract from his father's notes dated 1891, has no doubt that Jeremiah was born in Toxteth and he identifies Lower Lodge Farm, where the Otterspool stream joined the Mersey, as his birthplace. In the middle of the nineteenth century, the urban spread of Liverpool had still not reached the rural seclusion of this part of Toxteth:

> I have often seen Horrocks' birthplace and in the summer of 1859 I spent an afternoon and part of an evening in it, as some of my friends were then lodging there. It was a snug comfortable farmhouse, the road to it being from the narrow lane near Jericho Farm, which led to a well kept garden . . . There was a room on each side of the doorway and as we were

> in one of these rooms at tea time we were talking about its being the house in which Horrox was born, and my friends told me that in the kitchen there was nailed upon a beam an old brass plate stating – 'J. Horrox was born in this House'.[3]

The house was demolished in 1862, when a railway was built passing through the site, and a station called Otterspool was built on the land. Just before the house was demolished, an artist called Edward Cox made a sketch of the building. The authenticity of Lower Lodge Farm is supported by the story of an elderly labourer who remembered helping with the demolition and who confirmed the story of a worn brass plate attached to one of the beams. The description sounds authentic enough, but at least two other sites for the birthplace have been suggested in Toxteth: Baker's Shore Cottage and the 'Three Sixes house', so called because it carried the date 1666 above the doorway. Both these buildings were built well after the birth of Jeremiah Horrocks and their claim can be safely dismissed as being of little value. Thus Lower Lodge Farm may be identified as the birthplace with some degree of confidence; but there were plenty of happy guessers around in the nineteenth century, so it cannot be taken as a certainty. Jeremiah's father, for example, was confidently identified for many years as William Horrocks of Toxteth Park before later evidence proved that this assumption was quite wrong.

Strictly speaking, Toxteth Park did not become a part of Liverpool until the boundary extension of 1835, but it was so close that Toxteth relied on Liverpool for nearly all its trade and services. We have some excellent descriptions of seventeenth-century Liverpool from Celia Fiennes and Daniel Defoe, both of whom were greatly impressed by the trade and the growth of the town at that time, but their accounts were both written towards the end of the century, when trade with the American colonies had developed. A more accurate view of the Liverpool the young Jeremiah knew is given by William Camden, who visited in 1586. At that time, the name of the town was sometimes pronounced with only two syllables:

> The Mersey spreading and presently contracting its stream from Warrington falls into the ocean with a wide channel very convenient for

> trade, where opens to view Litherpole, commonly called Lirpoole, from a water extending like a pool, according to the common opinion, where is the most convenient and most frequented passage to Ireland: a town more famous for its beauty and populousness than for its antiquity; its name occurs in no ancient writer except that of Roger of Poictou who was lord, as stated of Lancaster, built a castle here, the custody of which has now for a long time belonged to the noble and knightly family of Molineux, whose chief seat is in the neighbourhood of Sefton, which Roger aforesaid in the early Norman times gave to Vivian de Molineaux. This Roger held, as appears in the Domesday book, all the lands between the rivers Ribble and Mersey.[4]

When Camden wrote that the town was famous for its 'populousness', he was guilty of some exaggeration. The Liverpool he saw had a population of less than a thousand and the number of residents was still in three figures by the time Jeremiah Horrocks was born. When Camden praised Liverpool's beauty, however, he was not guilty of any form of flattery. By the standards of the times, Liverpool was a very clean and attractive seaside town with fine beaches of golden sand. To the north were huge virgin sandhills which stretched nearly twenty miles along the coast, much further than the eye could see. Across the river lay the rural Cheshire peninsula of the Wirral, bounded by the Mersey and the Dee. The seaward vista showed the white-capped peaks of Snowdonia on the western horizon. To the north were the distant purple mountains of the Lake District and on a clear day Snaefell and the mountains of the Isle of Man could be seen from the higher vantage points.

The town itself contained only seven streets. It covered a rough semicircle, spreading less than a mile along the banks of the Mersey and expanding to about half that distance inland. The main street was the High Street, also known by its original name of Juggler Street, where street entertainers had performed since the Middle Ages. There were three streets radiating from each end of Juggler Street. At the northern end, Chapel Street ran down to the Mersey, and at the riverside was the chapel of St Nicholas and the little chapel of St Mary del Key which had been converted from a place of worship into the local school. There was no parish church in Liverpool. St Nicholas was no more than a chapel, and the town lay within the parish of Walton on the Hill, with the parish church about three miles away. In Tithebarn Street stood the barn that held the tithes from the produce of the town fields, an agricultural area to the north, laid out as strips in the Middle Ages and conveniently situated just outside the residential area. At the southern end of Juggler Street there was a square called 'High Cross' which boasted the village cross, the stocks and the pillory. Water Street ran from High Cross down to the riverside, where a fortified tower stood. Dale Street, named after a leafy dale about half a mile away just outside the town, ran inland from High Cross. The road crossed over the Townsend Bridge, where a stream carried the town's fresh water supply. The bridge was a great centre for gossip, where the women drew water above the bridge and did their washing below. It would have been well known to young Jeremiah because the Townsend Bridge was on the road to Toxteth Park. The continuation of the High Street to the south was called Castle Street and it led directly to the castle, the largest edifice in Liverpool, a great fortress of red sandstone that had dominated the town since it was founded in the time of King John. Lapping against the massive walls of the castle lay the tidal inlet known as the Liver Pool, 'Liver' being a corruption of the word 'lither', meaning a rushy pool. This small natural harbour was easily large enough to hold the shipping of the day.

Brownlow Hill, over the Townsend Bridge, was the place where the windmills set their sails to the wind to grind the corn. To the north was more high ground where the older generation remembered the Everton Beacon burning brightly in 1588 to warn the nation of the approach of the Spanish Armada. Liverpool was a lively and busy market town for south Lancashire and the Wirral area of Cheshire. On market days, fishermen arrived by boat from the villages along the coast to sell their catch. Ruddy-faced Cheshire farmers arrived on the ferries, large clumsy sailing boats which plied across the Mersey loaded with cattle and livestock. Rustic horse-drawn carts rumbled along the dusty road from the Lancashire plain, laden with the produce of the fields. At the Liver Pool there was a brisk maritime trade with places further afield, with Ireland, Scotland and Wales. There was some international trade, with ships from Europe importing wine and other luxury items. There was no trade with the American colonies. When Jeremiah was born, the Pilgrim Fathers had yet to set sail on the *Mayflower* and the Liverpool Atlantic trade did not begin to develop until the 1660s.

In Tudor times, Toxteth Park was a royal deer park covering an area of about ten square miles. It was bounded by the River Mersey on the west and, before Jeremiah Horrocks's time, it was fenced off and gated on the inland side to keep the royal deer from straying. In 1591 Toxteth was 'disparked' by the fourth earl of Derby and divided into twenty tenements, but it was still known as Toxteth 'Park'. A list surviving from 1596 gives the names of the twenty tenants, including Edward and Anne Aspinwall, showing that the Aspinwalls were amongst the first inhabitants of Toxteth Park. The town records show that Edward owned a water mill, fed by the Otterspool stream.[5]

There are no precocious childhood anecdotes about the young Jeremiah growing up in Toxteth. His early death and the outbreak of the Civil War postponed his fame until a time when there were few left alive to remember him as a boy. There is, however, plenty of information about the environment in which he was brought up and there are anecdotes about characters on both sides of his family. When it came to producing interesting characters, the Aspinwall family on Jeremiah's maternal side rivalled the Horrockses, and because Toxteth was Aspinwall country, they were a greater influence on young Jeremiah. Amongst the male Aspinwalls, the most common family names were Edward and Thomas, so that males with these names have often been confused by the genealogists. In the early 1600s, the senior member of the family was Edward Aspinwall, the uncle of Mary Aspinwall and therefore a great-uncle to Jeremiah. The Aspinwalls, like the Horrockses, were active Puritans and that is how the two families came to be so closely connected.

Before the rise of Cromwell during the English Civil War, the unorthodox beliefs of the Puritans excluded them from public office and they therefore tended to direct their efforts into trade and industry. Thus, by about 1600 the Aspinwalls were involved in the watchmaking trade and the quality of their watches was becoming known much further afield than Lancashire. The origins of watchmaking in England are obscure. Little is known about the various specialized trades that developed in the sixteenth century and how the artisans interacted nationally and with continental watchmakers. Unlike the cordwainer, who could make a pair of shoes with his own labour and a few simple tools, the watchmaker could not possibly manufacture all the parts needed for a watch. He depended on wire-drawers, spring-makers, wheel-grinders, engravers, silversmiths and sometime goldsmiths. Very few English watchmakers can be identified in the first decade of the seventeenth century, but one of these few is Thomas Aspinwall, the maternal grandfather of Jeremiah Horrocks. A watch of his manufacture, made in about 1606, survives at the Prescot Museum. It is an excellent early example of an oval watch, a fashion imported from Nuremberg. The oval watch became common in the following decades and watches in this style would therefore have been familiar objects of wonder to young Jeremiah. They were lavishly engraved and were popular with ladies, who wore them on the hip, attached to a chain. The watches had no minute hand, only an hour hand. They would run for twelve to sixteen hours and therefore had to be wound twice a day. They were not exceptionally accurate timekeepers, but they could be used to tell the time by touch at night:

> The dial is of silver, and has mounted thereon a brass hour ring. At each hour, near the exterior edge of the ring, is a slight knob to allow of the time being ascertained by feeling the hand and estimating its position with relation to the knobs. Over the hour ring is the engraved inscription, 'Our time doth passe a way'.

FIG. 1. Oval watch, *c.* 1606
Oval watch by Thomas Aspinwall, grandfather of Jeremiah Horrocks.

> The case is of silver. On the movement plate is engraved, 'Thomas Aspinwall fecit'.[6]

The family business was very successful. When Samuel Aspinwall, a first cousin to Jeremiah, died in 1672, the business had prospered so much that his estate was valued at £2,268 at a time when fifty pounds a year was seen by many as a reasonable income. Samuel owned premises in London and he had gained a considerable share of the London market. His sister Elizabeth had married Christopher Horrocks, also a cousin of Jeremiah and also a watchmaker. This second Horrocks–Aspinwall marriage underlines the close involvement and connections between the two families. Their reputation was such that in Dugdale's visitation of Lancashire[7] Susannah Aspinwall is mentioned as the daughter of 'Edward Aspinwall of Aspinwall, first watchmaker in England'. Whilst the sentiments expressed sound like a boastful exaggeration, the great success of the family watchmaking business is not in doubt and the Aspinwalls could have acquired the bulk of the London market by this time.

The watchmakers had already separated from the clockmakers as a trade. Whereas public clocks were fashioned from iron, watches were made from small parts of brass and steel, usually with precious metals for the case. The public clocks were maintained and repaired by the blacksmith, who could fashion the parts in his forge. Making and repairing the watches was very delicate work and it was far removed from the blacksmith's hammer and anvil. Watches were status symbols for the wealthy, with fine artistic engravings on the case and movement, and they were frequently decorated with gemstones.

The importance of an accurate timekeeper to the astronomer is obvious, but the accurate long-case pendulum clock belongs to the generation after Horrocks. In the early 1600s, the large public clocks and smaller lantern clocks for domestic use were all regulated by the foliot balance and were of limited use to the astronomer because of their poor timekeeping. The watches were even less accurate than the clocks, but they all had a link with astronomy because the timepieces, great and small, had to be regulated by reference to the Sun. Young

Jeremiah's first venture into astronomy was to measure the local noon using the sundial so that the watches could be set accurately to a fraction of an hour and regulated to keep good time.

The watches, even if they had been more accurate, were made to measure solar time, which was not the same as the time calculated from the stars. What the astronomer needed was sidereal time. He had to make an additional calculation to the solar time, or he had to have a special watch made for him. In practice, a good astronomer could tell the time far more accurately from the stars than from any clock or watch of the period, and it was sidereal time he wanted: solar time was an inconvenience to him. Nevertheless, the watchmaking trade did much for Jeremiah as an astronomer. The delicacy of the work explains why, when the boy came to study the heavens, he was able to measure the positions of the stars and planets with a watchmaker's precision, more accurately than anybody else before him.

The problem of Jeremiah's early education has to be addressed. Was it the watchmaking trade and the measurements on the sundial which first stimulated his interest in the heavens? Did his Puritanical upbringing stimulate him to study the origins of the world? Perhaps it was not until he entered the intellectual environment of Cambridge that he decided to become an astronomer. In the sixteenth century, the Liverpool Town Books record that 'wee agre that it is nedeful to have a lernyd man to be our schole master for the preferment of youth in this towne'.[8] John Ore, a learned man from London, came to take up the post on a salary of ten pounds per year. This local grammar school still existed when Jeremiah and his brother Jonas were of school age, but there are no surviving school records to indicate that they received their education at the school. There was another school, however, situated in Toxteth, which they could have attended. It was closely connected with the Aspinwall and Horrocks families, but there is evidence that it may have closed down in the very year that Jeremiah was born. The Toxteth school was the forerunner of the first place of worship in Toxteth Park. The history of Puritanism in Lancashire includes the following passage which places great emphasis on Edward Aspinwall. The account is written by a well-meaning Victorian and is therefore not as valuable as a contemporary account, but it gives us some idea of Edward's character and may be taken as fairly authentic. It throws some light on the situation in Toxteth just before Jeremiah was born:

> We meet with another group of Puritans, although their Puritanism seems to have been of a milder type than that of the moorlands, in and around Toxteth Park, near Liverpool. They erected a chapel in which they could hear the evangelical doctrines of the Reformation preached in their purity, and lift up a standard against the popery abounding in the neighbourhood. They invited Richard Mather when a boy to teach their children, and, when only youth, to teach themselves. Time has spared the name of one of them... Edward Aspinwall, the intimate friend of the sainted Mrs. Brettargh, and her comforter in the last hours of mortal sickness. The Church of Christ has some reason to venerate his memory, for by the influence of his holy conversation, his beautiful example, and his domestic piety the young schoolmaster, Richard Mather, was won over to the Puritan cause and prepared for the great work he did in New England. What the Mathers, father and four sons and many grandsons did for New England may, under God, be attributed in no small degree to the holy life of Edward Aspinwall.[9]

The writer introduces Richard Mather (pronounced with a hard 'a'), a well-known figure in the early history of the American colonies, a theme which keeps reappearing in Jeremiah Horrocks's early years. He was born in 1596 at Lowton in the parish of Winwick. He was educated at the local Winwick grammar school under a 'severe master' who talked his parents out of apprenticing him to a Catholic merchant. The name of the severe master was William Horrocks. The relationship between William and Jeremiah Horrocks is not known for certain, but the most likely possibility is that the schoolmaster was Jeremiah's uncle William, who died in 1618.

At the youthful age of fifteen, Richard Mather succeeded William Horrocks to become the master of his school at Winwick. He remained there one year, before moving to the household of Edward Aspinwall and setting up his own school at Toxteth. Edward Aspinwall obviously knew all about the promising youth at Winwick and managed to poach him away for his own purposes. The clue to the question of how Edward Aspinwall knew about Mather in the first place must lie with the schoolmaster William Horrocks and the family connections.

The Toxteth school was established in 1612. In spite of his youth, Mather was an excellent teacher. The school gained a good reputation and attracted pupils from miles around. But Richard Mather was still very young and he was not fully qualified. With help of his mentor Edward Aspinwall, he obtained a place to study at Brasenose College, Oxford and the college records show that he arrived there on 9 May 1618. Richard Mather had made such a good impression on the people of Toxteth that they wanted him to return from Oxford as their preacher, to instruct themselves and their children. The next development seems rather illogical, but Edward Aspinwall, who had gone to so much trouble and expense sending his prodigy to Oxford, called him back after just six months to run the chapel at Toxteth. Mather preached his first sermon at Toxteth on 30 November 1618 and in March 1619 he was ordained by Thomas Morton, the bishop of Chester.

Richard Mather, as a dedicated Puritan, rejected all forms of pomp and ceremony left over from the church's Catholic origins, and for fifteen years he preached his own interpretation of the gospel to his Toxteth congregation. The church authorities were prepared to tolerate Mather's unorthodoxy, almost certainly because they thought it would help their battle against the Catholics. Then came a crisis. The bishop discovered that Mather did not wear the mandatory surplice when he took the service. He was brought to trial before an ecclesiastical court, where he openly admitted the matter of the surplice. The bishop of Chester reinstated him but with a warning to conform and to wear his surplice at all times during the service. Richard Mather had no intention of changing his ways, and the incident led him to think about moving on from Toxteth to join the colonists in America, where he would be free to preach and act according to his own convictions.

The Toxteth residents appreciated that they had a good teacher in Richard Mather, and a fine preacher. He was so close to the Aspinwall and Horrocks families that Mather must have had a major influence on Jeremiah's early education. There is evidence that the Toxteth school continued to exist after Mather departed for Oxford. The name of George Wharledale appears in 1625 as a schoolmaster, and he probably taught the seven-year-old Jeremiah Horrocks.[10] Mather was seen to be of more use to the community as a preacher, running the chapel and educating adults in Puritanical principles, than as the master of the school, educating children. He probably did find some time to teach the youth of Toxteth, however. It was a small community and when Jeremiah Horrocks was a little older he must have received instruction from Richard Mather, though not necessarily in the formal environment of the schoolroom. Whatever the boy did learn at school, it is unlikely to have helped him with his astronomy, for there was little room in Mather's curriculum for anything other than religion. It was to the credit of the Puritans that they made the religion as simple as they could. They were anxious to 'purify' the faith and to get back as close as they could to the original Christian principles. Their views extended beyond their religion and they did not hold with witchcraft, magic or astrology. Thus, when Jeremiah came to apply his mind to astronomy, the logical and straightforward thinking of Richard Mather and the Toxteth Puritans stood him in good stead and he already had an inbred suspicion of astrology. The theology was also of benefit to Jeremiah in that the scriptures and the classics provided him with a far better university entrance qualification than the non-existent mathematics and science of the times. Edward Aspinwall and Richard Mather were both Oxford men and both were quite capable of teaching a bright pupil sufficient theology and Latin grammar to gain a place at university.

In 1624 Richard Mather married Katherine Holt. He remained at Toxteth until 1633 and was still running the chapel when Jeremiah Horrocks left home to go to university. Jeremiah did not enter Oxford as his two mentors had done. He followed the Horrocks tradition and went instead to Emmanuel College, Cambridge. He was not the first Horrocks to enter Emmanuel and there were other reasons why he chose the college. There were parts of the country where the Puritan movement had gained an even stronger hold than in Lancashire, and these included East Anglia and the university town of Cambridge. Nowhere in England was there more fervour for the Puritan movement than in the college where Jeremiah Horrocks had been offered a place to study. The son of James Horrocks and Mary Aspinwall was well briefed in Puritan theology. In 1632 the most Puritanical of the Cambridge colleges was quite happy to accept the watchmaker's son as one of their students.

CHAPTER THREE

Cambridge

It was the early summer of 1632 when Jeremiah Horrocks left his home in Toxteth to make the long, slow journey across England from the seaside town on the west coast to the fenlands of East Anglia and the ancient university town of Cambridge.

He would have travelled on horseback or possibly perched on the lumbering cart of a carrier and the journey would have taken him about one week. Jeremiah did not travel alone. He was accompanied by his second cousin Thomas Horrocks, a nephew of the nonconformist Revd Alexander of grey gown fame. Thomas was already a student at St John's College, Cambridge and was about two years older than Jeremiah. The branches of the Horrocks family were in touch with each other and there is little doubt that it was Thomas who first introduced his cousin to the life of the university. By 18 May Jeremiah had arrived in Cambridge and his name appears in the admissions register of Emmanuel College under the Latinized spelling 'Jeremy Horrox [of] Lancastre'. It is worth explaining here the confusion which arises from the two spellings of his name. Local records of the family, such as parish registers and wills, almost invariably use the spelling 'Horrocks', with 'Harrocks' as the most common variant. The spelling 'Horrox' is a Latinized version of the name, used later in his life for his Latin publications and also in some of the Cambridge records where Latin was the standard language. The county of each student was also recorded, so 'Lancastre' refers to Lancashire and not the county town of Lancaster. Thus the entry immediately before Horrocks was John Witham from Essex. Ralph Pook from Cheshire was the next student to be recorded.

When Jeremiah Horrocks first arrived at Cambridge, he was fourteen years old. In the 1630s, to enter Cambridge at this age was not the sign of a child genius. Schooling was usually completed by fourteen and the capable grammar-school boy was well versed in the Scriptures, Latin grammar and sometimes a smattering of Greek. There was no more to learn from the restrictive school curriculum and higher education usually meant the study of classics and theology. It has been suggested by his early biographers that Horrocks was born in 1619, which would make him only thirteen when he entered university. Under-age students were not uncommon, but they had to be recorded as such in the register, and since Jeremiah was not recorded as under age it may be assumed that he was born before July 1618.

On 5 July 1632 Jeremiah Horrocks's name was entered in the matriculation register. The freshmen were sworn in before the vice-chancellor, Thomas Comber, the Cambridge professor of Divinity, in the presence of the college masters, including William Sancroft the elder, who was both master and bursar of Emmanuel. Also present were proctors Turwhitt and Gatford, and James Tabor of the registry. This time, the name is spelled 'Jeremy Horrocks' and he appears amongst twenty Emmanuel freshmen in what was a fairly typical sample of three fellow commoners, ten pensioners and eight sizars. From the day they arrived, the students were divided into these three categories, representing a clear social division. The fellow commoners were of the upper class, the sons of wealthy and titled families, and paid double fees of about fifty pounds per year. The pensioners were so called because they paid for their pension, meaning their board and lodgings. They did not enjoy as many privileges as the fellow commoners, and were lower in the social scale, but most of them were the sons of minor gentry and clergy. The sizars were the lowest of the three categories. They had to wait on the fellows of the college to earn their keep, and had to run errands, empty the bedpans and perform other menial tasks. Jeremiah Horrocks was a sizar.

On the day when Jeremiah stood in the Cambridge registration queue, he struck up an acquaintance with another boy from the same college, and subsequent events show that the pair formed a youthful and lifelong friendship. His name was John Worthington, he was a sizar and he was also from Lancashire. Their names appear together in the university matriculation register, an indication that they were together in the queue and that they had befriended each other by July. John Worthington was not quite as fresh as his friend, having arrived at Emmanuel College on the last day of March, about seven weeks before Jeremiah. He had attended Oundle School and his father was Roger Worthington, a Manchester draper. Both boys befriended John Wallis (or Wallys), a pensioner of Emmanuel College who had already matriculated. Wallis was educated at Felsted School. His father had been the vicar of Ashford in Kent but died when John was six years old. He had been brought up by his mother and grandparents.

At this time Emmanuel College had existed for nearly half a century. It was founded in 1584 by Sir Walter Mildmay (1520–89). The college was a Puritanical Elizabethan foundation and as such, when Sir Walter appeared at the court of Queen Elizabeth, he found that his new college did not quite meet with her approval. According to an apocryphal story told by Fuller:

> Coming to Court after he had founded his Colledge, the Queen told him, 'Sir Walter I hear you have erected a Puritan Foundation.' 'No, Madam', saith he, 'farre be it from me to countenance any thing contrary to your established Lawes, but I have set an Acorn, which when it becomes an Oake, God alone knows what will be the fruit thereof.'[1]

Mildmay's predictions were well founded: many students went on to become masters of other colleges and some to even greater things. In the next century, when Fuller came to write the history of Cambridge University, he was able to say: 'Sure I am at this day it [Emmanuel] hath overshadowed all the University, more than a moiety [a half] of the present Masters of Colledges being bred therein.'

The first master of Emmanuel was an extraordinary character called Lawrence Chaderton, born in about 1536. In 1576 he married Cecily Culverwell, the daughter of a wealthy London haberdasher. He was forty-eight years old in 1584 when Emmanuel College was founded and he was still the master of Emmanuel thirty-eight years later, at the age of eighty-six. He then retired from the post of master but retained all his faculties and kept up his connections with the college. He was ninety-six when Jeremiah Horrocks arrived at Emmanuel as a freshman. Chaderton was a Lancashire man and loved to spend his vacations touring his native county, preaching the gospel wherever he could find an audience. It is probable that Jeremiah had already met Chaderton before his arrival in Cambridge, but even if this were not the case, he certainly knew something about this lively nonagenarian. Lawrence Chaderton was born to a family of Lancashire Catholics, but at some point in his youth he completely changed his religious code and became a fanatical Puritan. The Horrocks and Aspinwall families were so involved in the movement that they must have been aware of his ministries. They knew him pretty well and probably used his influence to gain places for their sons at Cambridge. As Chaderton approached his century, odes were already being composed for the funeral of the first master of Emmanuel:

> Wert thou ere young? For truth I hold,
> And do believe thou wert born old,
> There's none alive I'm sure can say
> They knew thee young, but always gray.[2]

The epilogists were premature. Lawrence Chaderton refused to die. In the library of Trinity College there is a book, *A sovereign antidote against Sabbatarian error*, published in 1636, in which, at the age of one hundred, the sharp-witted Chaderton scribbled the comments 'Who can prove this?', ' What's this?', 'Show examples'. He claimed to be 104 years old when he died in 1640. Some said he was only 102, but even if they are right it detracts very little from the story.

The popular view of the Puritans as psalm-singing killjoys who read nothing but the Bible is certainly true of a minority, but it is a great oversimplification. The picture is distorted by their detractors, the Cavaliers, who, during the Civil War, wished to present them in the worst possible light. When it came to university students, there was bound to be a youthful reaction against the continuous Bible-thumping to which they were subjected by their elders. In this respect there was little to choose between a Puritan college and the other colleges of the university. There was an initiation ceremony at Emmanuel College known as 'salting and tucking'. The freshmen were summoned to hall to meet with their seniors, where they were obliged to tell a joke or, in the wording of the times, 'pronounce a witticism'. If the audience laughed, the newcomer was accepted and rewarded with beer. If the humour was not appreciated, the fresher was 'salted': this meant he was obliged to drink a revolting salt-based concoction. Then came the 'tucking' ceremony, a painful and bloody process which involved making an incision in the lip and an abrasion from lip to chin.

The admonition book of Emmanuel College records some cases of extreme misbehaviour, such as when students slandered each other with libellous names, including 'puppy', 'rake hell' and 'varlet'. The college gates were locked at nightfall, but some revellers were happy to stay out all night, drinking and singing rude songs. The most common recorded offence was 'coming over the walls', and on more than one occasion this feat was accomplished with a ladder 'borrowed from Manning's wife'. Bear-baiting at the nearby village of Chesterton was an immensely popular diversion, as were the bawdy, lowbrow comedies written by the recently deceased playwright William Shakespeare. Stourbridge Fair and the midsummer games at the Gogmagog Hills were good excuses for the students to indulge in their pugilistic pleasures. They fought with fists, knives, cudgels and catapults, tore each other's gowns and ruffs, and had to put some of their colleagues under the pump to cool them off. At Candlemas it was the custom for some of them to let off fireworks with great abandon and no safety regulations. Women were not allowed in the college. Most of the students were still too young to fully appreciate the attractions of the fair sex, but this did not prevent the more mature students from enjoying some sexual liberties in the brothels of Cambridge. Among the more hygienic traditions was the reward of two pence for informing on scholars who passed water or emptied their chamber pots in unsuitable places.[3]

In the fifty years from 1596 to 1645, a total of 2,606 students entered Emmanuel College. In 1624 the input of freshers reached a peak at eighty-two, but by 1632, when Jeremiah Horrocks arrived on the scene, the intake had fallen a little. The college was therefore a small community which totalled about 250 members, including students, postgraduates, fellows and other academic staff. We know from his later correspondence that Jeremiah Horrocks had an outgoing and pleasant personality. It is therefore reasonable to assume that he knew everybody at his college, although he obviously had far more contact with his fellow sizars than with the college fellows and master.

During his time at Emmanuel, Jeremiah saw a few innovations. A new four-storey residential wing was built in the Dutch style, with dormer windows and curved gable ends. It became known as Old Court. He also saw the construction of a channel called Hobson's Conduit which brought a new fresh-water supply into the college. The works were supervised by Walter Frost, manciple of the college, whose job required him to deal with Thomas Hobson, a Cambridge tradesman who appears to have funded the conduit in his will. Hobson was the notorious carrier who hired out horses and carriages to town and gown, refusing to allow his customers a choice from his selection of steeds and only permitting them to take the next horse in rotation. He would not budge for love nor money. It was Hobson's choice: the next horse in line or no horse at all. He could not be bribed, and was so adamant with his rule that the saying 'Hobson's choice' entered the English language. Thomas Hobson died in 1630 at the age of eighty-seven, just before Horrocks arrived on the scene, but Jeremiah's cousin Thomas, and many of the older students and fellows could have told him plenty of stories about the recently deceased carrier.

At Cambridge and elsewhere, the Puritans were unhappy with the way the Church of England attempted to control the format of their services and their religion. The more adventurous among them had reached the conclusion that the only way to achieve their aims was to leave England altogether and set up a colony in the New World, in a place which became known as 'New England'. Thus it came about that in 1620 the Pilgrim Fathers made their momentous journey across the Atlantic in their tiny vessel, the *Mayflower*. This was a badly timed and in many ways foolhardy voyage. The pilgrims had an incompetent navigator, who guided them to the American coast but many miles south of their expected landfall. They followed the coast to the north and landed at Plymouth Rock in November, but had to wait many months, through the cold, harsh winter and half the summer, before they could reap their first harvest. On the face of it, the venture did not appear to be a well-planned attempt to found a colony, but in spite of the many deaths and hardships the Plymouth colony survived and gave encouragement to those who wanted to follow.

News of the Pilgrim Fathers was received with great interest at Cambridge and especially at Emmanuel College. The *Mayflower* venture had proved that it was possible to cross the Atlantic Ocean and to survive in the New World. After the voyage of the *Mayflower*, thirty to forty ships crossed the Atlantic every year. These vessels carried a few new settlers, but their main commerce was a return cargo of fish in what was little more than a development of the Newfoundland cod trade instigated by the Bristol fisherman at the end of the fifteenth century. They also traded with the Indians for furs and local produce. Thus in the 1620s major strides had been made in Virginia and New England towards the colonization of the new continent. In 1630 a new and very ambitious venture was planned by the East Anglian Puritans. They had somehow managed to obtain a royal charter allowing them to make a new settlement in Massachusetts.

In the summer of 1630, a brave little fleet of eleven ships sailed from Southampton to battle their way across the Atlantic and to round the choppy seas off Cape Cod. The flagship was the *Arbella*, named after the wife of the first governor of Massachusetts, John Winthrop from Groton in Suffolk, who was the leading light in the new venture. The pilgrims saw themselves as the pioneers of a new world. On the eve of their departure from England, it was the Revd John Cotton, husband of Elizabeth Horrocks, with his shock of white hair and the charismatic gestures of his right hand, who preached a moving farewell sermon to wish them well in their momentous undertaking.

Winthrop's fleet, which has a strong claim to be the first sent to colonize the New World, departed two years before Jeremiah Horrocks arrived in Cambridge, but throughout the 1630s and subsequent decades the migration continued and grew steadily. Emmanuel College was significant as the place where most of the early leaders and administrators of the New World were educated. The statistics show that of 129 Oxbridge men who migrated to America before 1650, no less then thirty-five were from Emmanuel College. Trinity College, Cambridge, with fifteen emigrants, was a poor second. When he arrived in Cambridge, Jeremiah Horrocks found himself right in the melting pot of colonial development. Everybody at Emmanuel was talking about America. He had landed amongst the greatest hotbed of Puritanical zeal in England.

The Revd John Cotton, the man who preached the farewell sermon for the Winthrop fleet, entered Cambridge via Trinity College but later moved to study at Emmanuel, where he became very involved with the Puritan movement. Born in 1585, he was older than most of the colonists, and his wife, Elizabeth Horrocks, was from the Turton branch of the Horrocks family (see the Horrocks family tree). John Cotton was the vicar of the parish church of St Botolph's in Boston, Lincolnshire, where he held the living for twenty-one years. During this period he became more and more Puritanical in his outlook, and after a time he ceased to observe certain Anglican religious rituals in the performance of his duties – a story which comes as no great surprise for we have come across it twice before. Emmanuel College was so impressed with Cotton's teachings that they frequently sent students to Boston, not just to hear him preach but to live with him for a time. Jeremiah Horrocks, because of his family connections, must have been one of these chosen students. Something of Revd Cotton's personality is suggested by the anecdotes which survive about him. One of Cotton's students wrote to ask his advice on playing cards, dancing and drawing names for valentines. He replied sharply that he did not approve of carding and valentines or of 'lascivious dauncinge to wanton dittyes'. There are times when one begins to wonder whether the Cavaliers had a point about the Puritans.

In 1631 Cotton's wife Elizabeth died and soon afterwards he

married his second wife, Sarah Story. In that year, the Church of England took legal action against him for not conforming to the formal service. Cotton's response, like that of Richard Mather and others of his creed, was to emigrate from Boston, England to Boston in America. In 1633 he sailed in the *Griffin*, accompanied by his wife and his lifelong friend Thomas Hooker. His status was very high amongst the Puritans and they lamented his departure. 'I saw the lord departing from England when Mr Hooker and Mr Cotton were gone, and I saw the hearts of most of the godly set and bent that way, and I did think I should feel many miseries if I stayed behind,' wrote Thomas Shepherd, one of his contemporaries. Sarah Cotton was heavily pregnant on her departure from England, a fact which makes one wonder whether 'foolhardy' might be a more appropriate adjective than 'brave' for some of the founders of the American colonies. Sarah gave birth to her first child on board a wooden sailing ship tossing on the mid-Atlantic swell. By some miracle, mother and child both survived and they called the baby Seaborn. The *Griffin* arrived safely at the Massachusetts Bay Colony, where John Cotton became an instructor at the First Church of Boston, a position he held until his death in 1652. The story is told that the peninsula of Shawmut, to which the colonists had moved in search of purer water, was renamed Boston, Massachusetts in his honour. Cotton's popularity in the colony was unbounded, and his influence in both civil and ecclesiastical affairs was probably greater than that of any other minister in New England. He wrote several valuable works, including his catechism, which he called *Milk for Babes, Drawn out of the Breasts of Both Testaments* (1646). It was widely used for many years in New England for the religious instruction of children.

Other emigrants among Horrocks's contemporaries at Emmanuel included Ezekiel Cheever, the son of a London spinner, educated at Christ's Hospital School. Cheever entered as a sizar the year after Horrocks and left for America in 1637. He became the most notorious teacher in the early history of Massachusetts: 'He held the rod for 70 years, the most famous schoolmaster in America'. He died, a nonagenarian, in 1708. Another, better-known contemporary was John Harvard, a mature student about eight years senior to Jeremiah but who was still at Emmanuel College in Horrocks's time. Harvard's father was a butcher and his mother was the thirteenth child of a cattle merchant who happened to be an alderman of Stratford-on-Avon. In 1625 the plague took John Harvard's father and most of his brothers and sisters, but John survived and went as a mature student to Emmanuel College. He took his BA in 1632 and his MA in 1635. John Harvard left no great mark on Cambridge, but in 1636 he married the sister of John Sadler, master of Magdalene College, and a year later he sailed for New England with his wife. Strictly speaking, John Harvard was not the founder of the college which bears his name and he did not obtain the charter which gave it corporate status. What he did provide was funds. He was already wealthy on his arrival in America and had the good fortune to inherit further wealth, which meant he found himself with money to spare. It was unfortunate for him that he did not enjoy his riches for long, since he died of tuberculosis in 1638 at the age of thirty, but his death benefited the Massachusetts colony. John Harvard left his substantial private library and half of his estate towards the foundation of a new college. At eight hundred pounds, this money was more than double the total sum scraped together by all the other colonists for this purpose. Six months after his death, the Great and General Court ordered that 'the college agreed upon formerly to bee built at Cambridge shalbee called Harvard College'. John Harvard never saw the college which bears his name.

During the long summer recess, when Jeremiah went home to visit his family in Toxteth, he was still exposed to a passion for the New World. At the chapel in Toxteth, Richard Mather, like John Cotton, had decided that he wanted to follow his own leanings towards Puritanism. He, too, sailed for America. Mather was unable to sail from Liverpool, the port from which the great majority of emigrants embarked in the nineteenth century. He had to make his way south to Bristol, whence he sailed on board the *James* on 4 June 1635. The Mather family arrived at Boston in the August of that year and Richard Mather's written journal of the voyage is one of the best accounts of an early crossing of the Atlantic. His reputation preceded him to New England, where he was offered several places in which to set up his ministry. He chose Dorchester, Massachusetts, where he remained until his death. Cotton and Mather became well acquainted. In 1656 Sarah Cotton and Richard Mather were both widowed. They secretly married each other and they had children. Later in the seventeenth century, Cotton Mather, the American Congregational minister and author, became one of the most celebrated of all the New England Puritans. He belongs to a later generation, but is worth mentioning since he had two Horrocks connections as the grandson of Sarah Cotton and Richard Mather.

With so many links through his college and family, Jeremiah Horrocks could not avoid the fervour and enthusiasm for the New World, and as a single young man about to embark on his career he was under great pressure from his peers. He must have given serious thought to joining the colonists. His cousin Thomas from St John's College did go to America, no doubt greatly influenced by his uncle John Cotton, and a few years later the cousins were writing to each other on astronomical matters. Jeremiah thus retained contact with his American cousin, but he had other ideas about the purpose of his own life. He knew that his ambitions could be attained more easily by staying in England. He was not convinced that Cambridge offered him the best environment for his chosen career but this did not lead him to consider emigrating. During his years as an undergraduate he had made up his mind to devote his life to astronomy and at this point it is at last possible to quote his own words to describe his feelings:

> There were many hindrances. The abstruse nature of the study, my inexperience, and want of means dispirited me. I was much pained not to have anyone to whom I could look for guidance, or indeed for the sympathy of companionship in my endeavours, and I was assailed by the languor and weariness which are inseparable from every great undertaking. What then was to be done? I could not make the pursuit an easy one, much less increase my fortune, and least of all, imbue others with a love for astronomy; and yet to complain of philosophy on account of its difficulties would be foolish and unworthy. I determined therefore that the tediousness of my study should be overcome by industry; my poverty

(failing a better method) by patience; and that instead of a master I would use astronomical books. Armed with these weapons I would contend successfully; and having heard of others acquiring knowledge without greater help, I would blush that anyone should be able to do more than I, always remembering that word of Virgil's

> '*Totidem nobis animaeque manusque*'
> [We have just as many souls and hands].[4]

Poverty by patience. Did he mean financial or intellectual poverty? Much has been written of his poverty, but it does not make a great deal of sense and it was probably invented by nineteenth-century philanthropists. It is true that he was a sizar, but this was because of a Puritanical frugality rather than real poverty, or he would not have been at Cambridge at all. The later phrases of the above extract seem to confirm that the context is intellectual poverty because he states that it can be overcome in time by patient study. It sometimes appears that Horrocks would have had more support if he had been at Oxford instead of Cambridge. There was no professor of astronomy at Cambridge, but at Oxford John Bainbridge became the first Savilian Professor of Astronomy in 1618. Oxford was therefore better endowed than Cambridge for astronomical studies, but in fact Bainbridge, Foster and Horrocks, the leading astronomers of the 1630s, were all educated at Emmanuel College. After Oxford, the only other centre of astronomical study in England was at Gresham College in London, where Henry Gellibrand was the professor of astronomy. In his later years at Cambridge, Horrocks communicated with Gellibrand, and when the latter was succeeded by the Emmanuel man Samuel Foster, Horrocks and Foster became regular correspondents.

It is difficult to support the claim that Jeremiah Horrocks knew what he wanted from life at the age of fourteen, before he arrived at Cambridge, and in spite of its limitations the university must be given most of the credit for instigating his interest in astronomy. There were some well-meaning people at Cambridge who were prepared to offer him support. One of these was Walter Frost, already mentioned in connection with Hobson's conduit. Frost was a great philosopher, but he was not very down to earth. He was said to have owned 'all manner of mathematical instruments' and he claimed to have invented a perpetual-motion machine. 'For by it the use of horses will be taken away . . . it may be made to flye throughout the air.' It is not likely that Jeremiah Horrocks, even as a student, was convinced by Walter Frost's claims of perpetual motion.

Emmanuel College boasted a fine library of over six hundred books. We have a very good idea of the contents of the college library at this time because six inventories survive covering the years from 1597 to 1637. The earliest of these inventories shows that in the first fifteen years of its existence the library grew from humble beginnings to 449 volumes. By 1610 there were 503 volumes, growing to 533 by 1621. It was considered to be a large and well-equipped library for the times, but many of the theological books were purged in 1635 – removed from the shelves because they were considered to hold papist opinions.[5] The library furnishings included a terrestrial globe with a wainscot frame and a portrait of Sir Walter Mildmay. The founder's picture, 'with a curtayne of blew saye & an iron rod', looked benignly down on readers. The library had nine glass windows and three casement windows. There were nine oak desks, each with three degrees, this being a term for the number of shelves in the desk rather than the angle to the horizontal. Some of the desks were used for storing the books. The practice of chaining the books to the cases to prevent theft was continued in some colleges until much later in the seventeenth century, and with good reason, for volumes were always going missing. Usually the books were simply stolen by unscrupulous students. Emmanuel was an enlightened college in as much as the books were not chained to the desks and the students seem to have been given good access to them. The college records state that a lock was put on the library door and offenders were given a stiff fine of 40 shillings for not returning books on time.

The theological works, comprising over four hundred volumes, accounted for the lion's share of the collection, with two-thirds of the total books. Preaching and collections of sermons comprised another twenty-five books. Science, astronomy and mathematics were not quite so well endowed, and the total number of books in these categories failed to reach double figures, despite the fact that the founder, Sir Walter Mildmay, had made a generous donation in his time as master: the number of scientific books still mustered only about 4 per cent of the total. History accounted for about sixty volumes, lexicology and philology about forty, and the Greek and Latin poets and orators were covered by about thirty books. Law and logic took up 2 and 1 per cent respectively of the library's holdings. Euclid's *Elements* and Ptolemy's *Almagest* were amongst the few mathematical works available, and Walter Mildmay's donation included a copy of *Institutiones Geometricae* (Paris 1535) by Albrecht Dürer, a reminder of the contribution to perspective geometry made by the great artists of the Renaissance.[6]

Jeremiah Horrocks was not the only student to complain about the lack of facilities in his chosen subject. His friend John Wallis grumbled about the lack of textbooks in mathematics, in a passage which echoes the feelings of his friend:

> I did thenceforth prosecute it [mathematics], (at School and in the University) not as a formal study, but as a pleasing Diversion, at spare hours; as books of Arithmetick, or others Mathematical fel occasionally in my way. For I had none to direct me, what books to read, or what to seek, or what methods to proceed. For mathematics, (at that time with us) were scarce looked upon as Academicall studies, but rather mechanical; as the business of Traders, Merchants, Seamen Carpenters, Surveyors of Lands, or the like; and perhaps some Almanac-makers in London. And amongst more than Two hundred Students (at that time) in our College, I do not know of any Two (perhaps not any) who had more of Mathematics than I, (if so much) which was then but little; And but very few, in that whole University. For the study of Mathematics was at that time more cultivated in London than in the Universities.[7]

The mention of London is a reference to Gresham College, founded in 1598 as a very different establishment from Oxford and Cambridge. Gresham College had professors but no undergraduate students. It offered public lectures on a wide variety of subjects, including mathematics and astronomy but not classics or theology. It was founded with the object of advancing useful knowledge such as navigation and manufacturing processes.

The classical section of the Emmanuel College library included Horace and Homer, but incredibly Virgil seems to have been ignored. The college could not possibly have disapproved of his works on theological grounds and in his later treatises Horrocks himself sometimes quoted from Virgil. At one time the library owned a copy of Chaucer but this was recorded as missing in 1632. There was nothing as modern as Shakespeare: his works were becoming popular, but were too racy for the older generation of a Puritanical college to give their approval. Later in his career, however, when we come to Horrocks's eulogies, we find that Shakespeare and the Elizabethan poets had made a great impression on him. Another glaring omission was the work of Francis Bacon. Bacon's works were published in the 1620s and they seem to have been well known to Jeremiah Horrocks early in his career: he quotes from Bacon and follows his codes of practice very faithfully. There seem to be other anomalies in Horrocks's university education. He managed to acquire knowledge which could not be found in the library at Emmanuel, but an enterprising student would search everywhere for information and he could easily have gained access to other libraries in the university.

Francis Bacon was born in 1561 and died in 1626. He was a Londoner, but he also knew something about Horrocks's part of the country, for he represented Liverpool in parliament from 1588 to 1594. It would be nice to think that Bacon knew men of the older generation like Edward and Thomas Aspinwall, but this is pure speculation and there is no evidence that Bacon ever went to visit his constituency. Bacon was very different from the majority of his peers. He was greatly influenced by practical men such as Caxton and Columbus. His *Novum Organum* (New Methodology) was published in 1620. The frontispiece shows a ship sailing through the Pillars of Hercules at the Straits of Gibraltar into the Atlantic Ocean in search of a new world. He was aware of the potential of America but did not live long enough to become involved with the great emigration fervour of the 1630s. The voyages of discovery, the practical inventions and improvements made before and during his lifetime greatly impressed Bacon. Examples were the invention of printing, gunpowder and the magnetic compass. It was his view that this knowledge had changed the whole face and state of things throughout the world. He was particularly impressed by the telescope and the discoveries of his contemporary Galileo Galilei. Bacon wanted to become involved with practical inventions and in the discovery of new worlds, by which he did not mean America or the other planets, but a new intellectual world. He diagnosed the defects of contemporary learning, and he suggested a plan of research which he hoped would lead to real knowledge and practical results. 'I publish and set forth these conjectures of mine just as Columbus acted, before that wonderful voyage of his across the Atlantic, when he gave the reasons for his conviction that new lands and continents might be discovered.' Bacon's ambition was to be the Christopher Columbus of the new intellectual world. The defect of traditional learning, such as he found in the old universities, seemed to him to be that of the medieval monks in their cells:

> This kind of degenerate learning did chiefly reign amongst the Schoolmen: who, having sharp and strong wits and abundance of leisure, and small variety of reading, but their wits being shut up in the cells of a few authors (chiefly Aristotle, their dictator), as their persons were shut up in the cells of monasteries and colleges, and knowing little history, either of nature or time, did out of no great quantity of matter and infinite agitation of wit, spin out unto us those laborious webs of learning which are extant in their books. For the wit and mind of man, if it work upon matter, which is the contemplation of the creatures of God, worketh according to the stuff, and is limited thereby; but if it work upon itself, as the spider worketh his web, then it is endless, and brings forth, indeed, cobwebs of learning admirable for the fineness of thread and work, but of no substance or profit . . .[8]

Bacon formulated a code of scientific procedure that would enable any sensible person to make scientific discoveries. His *Novum Organum* was intended as an instrument with which to achieve this. Just as a pair of compasses enables even an unskilled person to draw a good circle, so the new method should enable ordinary people to make scientific discoveries. Bacon, however, grossly underrated the place of originality and sagacity in the work of science, and the difficulties involved in extracting the laws of nature from experiments. He found it impossible to apply his elaborate rules to his own discoveries. In spite of this, a considerable amount of credit is due to him for his analysis of the scientific method. Horrocks read and absorbed Bacon's works, but he did not believe all of his conclusions. Astronomically speaking, Francis Bacon was still in the past, because he could not accept any theory which did not put the Earth at the centre of the universe.

In 1634, when Jeremiah returned to college after the summer recess for his third year of study, it was in the company of Edward Aspinwall. This was not his great-uncle Edward but the latter's grandson. Edward junior entered Emmanuel College as a pensioner, against the Aspinwall family tradition of going to Oxford. Richard Mather and the Aspinwalls were obviously very impressed with the developments at Emmanuel College and they were prepared to change their university allegiance to take advantage of the movement. Edward Aspinwall junior went on to a very successful career, and married Eleanor Ireland, a wealthy heiress and the daughter of Sir John Ireland. Later in the century, he was described in Dugdale's visitation as 'the first watchmaker in England'.

Amongst Jeremiah Horrocks's friends, John Worthington gained his MA in 1639 and in spite of his lowly start as a sizar became a fellow of Emmanuel College in 1642. In 1650 he became the master of Jesus College, Cambridge. Three years later, he became the rector of Gravely in Cambridge and held the living of Fen Ditton, Cambridgeshire from 1654 to 1663. He became the vice-chancellor of Cambridge University for the year 1657–8. He then held livings at Barking in Essex, at Needham in Suffolk and at Moulton All Saints in Norfolk. He became a lecturer at St Benet Fink's in London, then moved on to become rector of Ingoldsby in Lincolnshire and prebendary of Lincoln in 1668–71. He published religious writings, including an edition of the works of John Mede. He died at Hackney in 1671 and was buried there.

John Wallis had an equally brilliant career. He lived a long and active life and was still alive at the start of the eighteenth century. He obtained his MA in 1630 and moved to London. During the Civil

War he was employed by Parliament to decipher coded messages from the Royalists. He became Savilian Professor of Geometry at Oxford, where he became involved with John Wilkins and Christopher Wren in the foundation of the Oxford Scientific Club. He held his Oxford professorship from 1659 until his death in 1703 and he was the keeper of the archives for most of that time. He became an active member of the Royal Society and in 1655 published *Arithmetica Infinitorum*, which contained some of the seeds of the differential calculus, later developed by Newton and Leibniz. In the 1670s he edited *Opera Posthuma*, which will be explained in more detail later in the story. He invented the mathematical symbol for infinity. In 1690, in his seventies, he worked on the deciphering problem again, but this time he was on the side of the king, William of Orange.

The statistics for Emmanuel over the first half-century of its existence show that just over a quarter of those who can be identified as sons of the gentry did not even get as far as the matriculation ceremony in their first weeks at Cambridge. Over 70 per cent of them failed to graduate. The notion that Cambridge was little more than a finishing school for the wealthy is not without foundation, but it was not only fellow commoners and pensioners who left without a degree: sometimes the sizars failed to graduate. What happened to Jeremiah Horrocks whilst his brilliant friends went on to long careers and great things? In common with two of his fellow astronomers, Tycho Brahe and Galileo, he left university without a degree. He may have failed his exams and it would not be the first time that this had happened to a student who spent all his time studying subjects which lay outside the curriculum. It could have been that his parents withdrew him due to poverty, but there is no other evidence to support the theory that they fell on hard times. It could be that, in spite of being a devout Christian, he had no intention of entering holy orders and a degree was therefore of limited use to him. He knew exactly what he wanted to do with his life:

> It seemed to me that nothing could be more noble than to contemplate the manifold wisdom of my creator, as displayed amidst such glorious

works; nothing more delightful than to view them no longer with the gaze of vulgar admiration, but with a desire to know their causes, and to feed upon the beauty by a more careful examination of their mechanism.[9]

CHAPTER FOUR

Astronomy before Horrocks

At the time Jeremiah Horrocks resolved to study astronomy, the great majority of people still believed that the Earth was the centre of the universe. Radical new theories of the universe had been formulated and published, but in the early decades of the seventeenth century these ideas could take generations to become accepted. The church authorities, and the Roman church in particular, were opposed to any theory which contradicted the account in the Bible. But a revolution had already been set in motion and the invention of the telescope meant that new discoveries were being made in the heavens which showed that the ancient view of the universe was full of contradictions.

Since the dawn of history, every civilization has seen men who studied the skies. There were astronomers in Egypt, in Babylon and in the ancient civilizations of India and China. The knowledge was not confined to Europe and Asia. In America, the Incas and the Aztecs built pyramids and temples that showed their knowledge and fascination with the Sun, the Moon and the stars they saw in the night sky.

The commonly-held view of the universe was of a flat Earth at the centre of the world, with the Sun, Moon and stars revolving on a dome around it. Pythagoras, famous for his theorem about the square on the hypotenuse, lived around 580 BC and is thought to have been the first person to put forward the idea that the Earth was a sphere. For a short time his idea prevailed, but after his death it was soon forgotten. In the third century before Christ, Aristarchus of Samos revived the theory that the diurnal motion of the stars could be best explained in terms of the Earth as a sphere revolving on its axis and he also suggested the very futuristic idea that the Earth orbited around the Sun. His cosmological ideas were so advanced that they were quickly dismissed as impossible and heretical. It was nearly two thousand years before they appeared again.

Not long after Aristarchus, Eratosthenes of Cyrene (*circa* 276 to 194 BC), measured the size of the Earth and arrived at the very acceptable figure of 252,000 *stadia* (about 46,000 km) for the Earth's circumference. The Greek unit of the *stade* varied throughout the Hellenistic world and we cannot therefore be sure of his accuracy. Well-meaning researchers from later centuries have calculated the length of the *stade* from the circumference of the Earth, so it is not surprising to find that Eratosthenes' calculation was so exact. The important point is that his method, which involved the measurement of the Sun's altitude at two different latitudes, was scientifically very sound.

Hipparchus was one of the greatest astronomers of the ancient world. He was born in Nicaea, Asia Minor, early in the second century before Christ. We know that he moved from Asia Minor to Rhodes and that he was making observations at Rhodes from 141 to127 BC. We also know that he had access to data such as solar and lunar eclipses collected from the time of Nabonasser (*circa* 747 BC), more than six hundred years before his time. Hipparchus was the first astronomer to appreciate the vast scale of the solar system and amongst his achievements was the first accurate estimate of the distance to the Moon. He knew that on 14 March 190 BC, before he was born, there had been a total eclipse of the Sun at the Hellespont near Constantinople. The same eclipse had been observed in Egypt at Alexandria, but at ten degrees to the south the eclipse was not total and a maximum of only four-fifths of the Sun was covered by the Moon. By making the reasonable assumption that the Sun was very distant compared to the Moon, Hipparchus was able to calculate the Moon's distance, arriving at a figure of sixty times the Earth's radius. This was an excellent estimate for the very sound reason that Hipparchus' reasoning was perfectly correct.

The most amazing of Hipparchus' discoveries was that the Earth's axis did not remain in a fixed direction relative to the stars. He was able to show that it progressed around a cone, rather like the axis of a spinning top. Over a period of several thousand years, the pole traced out a circle in the sky. He calculated that every year the Earth's axis precessed by at least a hundredth part of a degree. This phenomenon is known as the precession of the equinoxes and there is some evidence that it was a rediscovery: it seems to have been known in ancient Babylon long before the time of Hipparchus. His other great discovery was the length of the year. He estimated that the solar year of 365 1/4 days was too long by about 1/300th part of a day. He therefore realized that over a period of three centuries a calendar with one leap year every four years would be one day or more in error.

The greatest astronomer of the ancient world is assumed by many to be Claudius Ptolemaeus of Alexandria (*circa* AD 100–170). His greatest work, the *Almagest*, was published in thirteen volumes. It showed how to calculate the positions of the planets and how to predict eclipses of the Sun and the Moon. It was the bible of astronomy for fifteen centuries. As Europe degenerated into the dark ages, it was fortunate that the work of Ptolemy was copied and preserved by Persian astronomers. The title *Almagest* is an Arabic word meaning 'the great work'. Ptolemy published a catalogue of 1,022 stars in forty-eight constellations. He gave their latitudes and longitudes on the celestial sphere and also a brightness scale from one to six. Many of the stars in Ptolemy's catalogue were not even visible from Alexandria, where he worked, and it is clear that he copied much of his data from the earlier star catalogue of Hipparchus. There has been a fashion for deriding Ptolemy as no more than a plagiarist, but this is an oversimplification. In Ptolemy's time, Hipparchus had been dead for nearly two hundred years and Ptolemy does in fact give full credit to the work of his predecessor. There are some instances where he gives us information about Hipparchus which would have been lost without his acknowledgments.

Only two co-ordinates are required to specify the position of a star in the night sky, but to identify a star from the catalogue the observer needs to know his own latitude and the sidereal time. The sidereal time gives the position of the rotation of the Earth with respect to the stars (as opposed to the solar time which is based on the position with respect to the Sun). Thus the motion of any star through the sky would be a perfect circle if it were not for the effects of atmospheric refraction, a small error which can easily be corrected. From earliest times, however, astronomers noticed that a small number of stars – just five – did not observe the perfect circle law. These wandering stars were called planets and are now known in English by the names Mercury, Venus, Mars, Jupiter and Saturn. The Sun and the Moon were also wanderers, making a total of the mystic number of seven bodies which wandered around the sky. The Sun was never visible against the starry background, except in the very special case of a total eclipse, but the ancients knew that, had the stars been visible, then they would see the Sun, like the Moon, moving against a starry background.

Nothing could be more obvious to the ancients than that the Sun dictated the hours of day and night. The position of the Sun in the zodiac determined the seasons, told them when to sow seed and when to harvest. The Moon also had some purpose in the scheme of the world. The tides were not great in the Mediterranean, but it was well known that the Moon had some influence over the sea and that the position of the Moon somehow governed the tides. In Egypt the astronomer priests knew that the day of the first rising of the Dog Star before dawn predicted the flooding of the River Nile. It was a logical deduction that all the bodies in the heavens had some meaning and that they influenced an earthly event. It was the purpose of astrologer priests to find the influence of all the bodies in the heavens.

The motion of the Sun through the zodiac seemed simple enough, but the motion of the Moon was very complex. Both of these motions needed to be solved in order to be able to forecast an eclipse. The Moon's motion was never fully solved in the ancient world, but certain regular cycles were discovered and as early as the time of Hipparchus the astronomers were able to predict an eclipse very accurately. The path of the Moon through the zodiac is only six degrees away from the elliptic. This means that the latitude of the Moon is not difficult to model, and, combined with a theory to predict the longitude of the Moon, it was possible to predict eclipses quite accurately. Hipparchus and Ptolemy each produced a theory to foretell an eclipse. Both theories were obviously wrong in one respect, because they predicted that the Moon's distance from the Earth varied by a factor of four throughout her orbit and a Moon four times its normal size would be very obvious to the observer. The theory was quite capable of predicting the date of an eclipse, however, and this meant that it served its purpose.

The planets moved in the same general direction as the stars as they crossed the night sky, but they also experienced a retrograde motion at certain times of year. After a period of time, the motion of the planet repeated itself and its position became predictable, but the problem for the astrologers was to devise a model for calculating the positions of the planets at any time in the future. Ptolemy was the first to define a system to solve this problem. It was based on a combination of two circular motions, a cycle to simulate the planet's motion around the Earth and an epicycle to simulate the retrograde motion of the planet. Thus in the Ptolemaic system the stars and the planets were carried around the Earth by a set of crystal spheres. The Moon revolved around the Earth on the innermost of the crystal spheres. Outside the sphere of the Moon lay the sphere which carried the Sun, rotating once per day. Outside the sphere of the Sun were five other concentric spheres, one for each of the five planets. The outermost and most distant sphere carried the stars, and rotated slightly faster than the Sun, making one extra rotation every year. The spheres carrying the planets moved in a more complex fashion, with cycles and epicycles. There were errors in the Ptolemaic system, but these were very small and difficult to detect with the instruments of the time. The astrologers were delighted with the system. It was excellent for casting horoscopes and for predicting the positions and conjunctions of the planets.

For a thousand years, whilst earthly empires rose and fell, the crystal spheres revolved in unison with the stars and the planets. In heaven, the angels turned the handles to keep the system in motion. The majestic harmony of the heavens was accompanied by the music of the spheres. Nobody challenged the system of the heavens. It was based on the perfect figure of the circle.

The new millennium arrived, another five hundred years passed by and it seemed that the angels would turn the heavenly spheres until the Day of Judgement arrived. Then, in 1473, a man was born at Torun in Poland who was destined to disturb the harmony. He studied at the University of Cracow and he became interested in the heavens. His name was Nicholas Copernicus. He used a parallactic instrument to study the Moon, a quadrant for the Sun, and an astrolabe or armillary sphere to measure the stars. As he calculated the orbits of the planets, it seemed to Copernicus that there must be a simpler model for the universe. If the Earth rotated about its axis, it would save the angels the perpetual labour of turning the celestial spheres. He took the idea further. If the Earth orbited around the Sun instead of the Sun orbiting around the Earth, the motions of the planets would be greatly simplified. He proposed a system whereby the Sun was at the centre of the universe, with all the planets, including the Earth, orbiting around it. It was obvious from the motions of Mercury and Venus that they were closer to the Sun than the Earth, but the outer planets of Mars, Jupiter and Saturn were further away. The theory relegated the Earth to be a body less than the Sun, but it explained why the planets were so much fainter at the distant points of their orbits, something that the Ptolemaic sysytem did not explain.

Ptolemy had known about the work of Aristarchus, who had suggested a system almost identical to that of Copernicus. Ptolemy had considered a moving Earth but had rejected the idea on common-sense grounds:

> [the motion] would have to be exceedingly violent and its speed unsurpassable to carry the entire circumference of the Earth around in twenty-four hours. But things which undergo an abrupt rotation seem utterly unsuited to gather bodies to themselves, and seem more likely, if they have been produced by combination, to fly apart unless they are held together by some bond. The Earth would long ago . . . have burst asunder . . . and dropped out of the skies.[1]

Copernicus's response was that rotation was a natural motion:

> What is in accordance with nature produces effects contrary to those resulting from violence. For, things to which force or violence is applied must disintegrate and cannot long endure, whereas that which is brought into existence by nature is well ordered and preserved in its best state. Therefore Ptolemy has no cause to fear that the Earth and everything will be disrupted by a rotation created through nature's handiwork, which is quite different form what art or human intelligence can contrive.[2]

The notion that the Earth was a sphere was generally accepted in the time of Copernicus. Ships could be seen disappearing below the horizon. The world had been circumnavigated by Magellan's ships and some of his sailors had returned to tell the tale. Contrary to popular belief, the Flat-Earth Theory had been superseded by astronomers long before Columbus, but this still left the Earth firmly at the centre of the universe. Copernicus considered the motion of a sphere in space and the effects of a diurnal rotation:

> Since it has already been proved that the Earth has the shape of a sphere, I insist that we must investigate whether from its form can be deduced a motion, and what place the Earth occupies in the universe. Without this knowledge no certain computation can be made for the phenomena occurring in the heavens. To be sure, the great majority of writers agree that the Earth is at rest in the centre of the universe, so that they consider it unbelievable and even ridiculous to suppose the contrary. Yet when one weighs the matter carefully, it will be seen that this question is not yet disposed of, and for that reason is by no means to be considered unimportant ... Now the Earth is the place from which we observe the revolution of the heavens and where it is displayed to our eyes. Therefore, if the Earth should possess any motion, the latter would be noticeable in everything that is situated outside of it, but in the opposite direction, just as if everything were travelling past the Earth. And of this nature is, above all, the daily revolution. For this motion seems to embrace the whole world, in fact everything that is outside of the Earth, with the single exception of the Earth itself. But if one should admit that the heavens possess none of this motion, but that the Earth rotates from west to east; and if one should consider this seriously with respect to the

> seeming rising and setting of the Sun, of the Moon and the stars; then one would find that this is actually true.[3]

Copernicus knew that his ideas would be seen as heresy, not only by the Roman church but by the Protestants as well. 'Mention has been made of some new astrologer,' said Martin Luther cynically,

> who wants to teach that the Earth moves around, not the firmament or heavens, the Sun and Moon. Just as though a man were to sit in a cart or a moving boat, and thinks he sits motionless and rests while the Earth and the trees move . . . This fool seeks to overturn the whole art of astronomy. But as the holy scriptures show, Jehovah ordered the Sun, not the Earth, to stand still.

Copernicus's work was published in 1543, the year in which he died. As he lay on his deathbed, his work was ready for the press. He did not live to see it accepted, but wisely delayed publication until after his death.

In the sixteenth century, new knowledge took a long time to circulate and even longer to become accepted. Very few read Copernicus's book, and of those who did, even fewer took his ideas seriously. The idea of a heliocentric universe would have died a death were it not for subsequent events. In 1546, three years after Copernicus died, a Danish nobleman called Tycho Brahe was born at Skane in Denmark (now in Sweden). Tycho studied at the Lutherian College in Copenhagen and was sent by his uncle to the University at Leipzig to study law. This course of study did not appeal to him, but Tycho was wealthy enough not to have to worry about getting a degree. What he really wanted was to study the skies. Tycho Brahe was one of the few people who had actually studied Copernicus. He accepted the Copernican universe and declared his own heretical ideas by rejecting the literal notion of the crystal spheres:

> There really are not any spheres in the heavens . . . Those which have been devised by the expert to save the appearances exist only in the imagination, for the purpose of enabling the mind to conceive the motion which the heavenly bodies trace in their course, and by the aid of geometry, to determine the motion numerically through the use of arithmetic.[4]

In December 1566, when he had just passed his twentieth birthday, Tycho quarrelled with a Danish nobleman called Manderup Parbsjerg. He came close to losing his life when he was challenged to a duel. Parbsjerg got the better of the duel and cut off part of Tycho's nose. The quarrel was quickly made up, but for the rest of his life Tycho's appearance was rather startling, as the bridge of his nose was made of gold, copper and beeswax. Fortunately for him, his injured nose did not affect his eyesight or his astronomical observations.

In 1572 he made a striking discovery. A new star appeared in the heavens and for several days was the brightest object in the night sky. It grew to be as bright as Venus and it was even visible after the Sun had risen. The star had no right to be there. Everybody knew that the heavens were fixed and unchanged for all time. He described the star as follows:

> In the evening, after sunset, when, according to my habit, I was contemplating the stars in a clear sky, I noticed that a new and unusual star, surpassing all the other stars in brilliancy, was shining almost directly above my head; and since I had, almost from boyhood, known all the stars of the heavens perfectly (there is no great difficulty in attaining that knowledge), it was quite evident to me that there had never before been any star in that place in the sky, even the smallest, to say nothing of a star so conspicuously bright as this. I was so astonished at this sight that I was not ashamed to doubt the trustworthiness of my own eyes. But when I observed that others, too, on having the place pointed out to them, could see that there really was a star there, I had no further doubts. A miracle indeed, either the greatest of all that have occurred in the whole range of nature since the beginning of the world, or one certainly that is to be classed with those attested by the Holy Oracles.[5]

At first the star was white, but it changed to yellow and began to grow fainter. By April, it was only a second magnitude star and he described it as 'lead coloured'. Tycho's new star, his obsession with astronomy and his skill in observing came to the notice of King Frederick II of Denmark. The king persuaded Tycho to give lectures on astronomy in Copenhagen. Tycho's knowledge and skill impressed Frederick and others. In 1576 Tycho received a royal summons from the king and was offered funding for an observatory at Uraniborg on the island of Hven in the straits between Denmark and Sweden. Thus Tycho became the master of Hven, a bleak and isolated kingdom of about two thousand acres. Mapping the skies was a long and lonely vigil for Tycho, but the observatory was built in a grand style, with running water in the bedrooms and the best instruments of the times for the study of the stars. For more than twenty years, Tycho and his assistants measured the positions of the stars and the planets from their island observatory. Tycho created the first star catalogue for over a thousand years – a rival to that of Ptolemy. He also measured the planets and carefully plotted their positions in the sky for twenty years. His catalogue had 777 fixed stars, and this was later increased to a total of one thousand by the less careful determination of 223 additional stars. Tycho's tables were eventually published in 1601, the year of his death, by his assistant Longomontanus.

When King Frederick died in 1588, the future of the observatory at Hven came under threat. Tycho Brahe managed to get on the wrong side of the new king, Christian IV, and his source of revenue was cut off. He had collected a great volume of data which he wanted to examine, but he was obliged to work his way around Europe to find a new sponsor. Eventually he found a patron in Rudolph II, the Holy Roman Emperor, a morbid and incompetent man who was very interested in astrology. Tycho then began to search for a mathematician to try to make some sense out of his data on the planets.

Tycho was the greatest and last of the naked-eye observers, but when it came to philosophy, his idea of the universe was a retrograde step after Copernicus. He thought that the heliocentric system of Copernicus was a good model, but he could not bring himself to move the Earth from its privileged position at the centre. Instead, he devised a system whereby only the Sun and the Moon orbited the Earth, but the other five planets orbited around the Sun. Mathematically, the model was almost identical to that of Copernicus, the only difference being that everything was referred back to the Earth as the stationary point. But it was a clumsy model. Tycho was not a great theorist and his name would not be held in such high esteem had it not been for the German assistant he employed towards the end of his life to help with his catalogue of the stars. His earlier assistant, Longomontanus, was working to try and improve on the motion of the planet Mars, but the problem was very complex and he was making very slow progress. The observations were given instead to Johannes Kepler. It proved to be a most fortunate choice.

Johannes Kepler (1571–1630) was born at Weil de Stadt in Germany. He was a small, frail man, near-sighted and plagued by fevers and stomach ailments. He compared himself to a snappish little house dog who tried to win his master's favour but in doing so drove others away. He was a strange and mystical character, very interested in astrology, and at least eight hundred of his horoscopes are still preserved. When he was casting horoscopes for his family, he described his grandfather as 'quick tempered and obstinate', his grandmother as 'clever, deceitful, blazing with hatred, the queen of busybodies', his father Heinrich as 'criminally inclined, quarrelsome, liable to a bad end' and his mother as 'thin, garrulous and bad-tempered'. In later life, he spent many months trying to clear his meddlesome mother of a charge of witchcraft. In 1597 he married Barbara Muller, who had already been twice widowed. When he first met her, she 'set his heart on fire', but unfortunately when they married the planets were in the wrong signs of the zodiac. Kepler became disillusioned with his wife when he found that she did not believe in his precious astrology. They had two children, but both died very young. He ended up accusing her of being 'fat, confused, and simple-minded'.

In spite of his personal problems, Kepler was an excellent mathematician and he was always looking for mathematical patterns in the universe. When he came to study the planets, he tried to fit regular plane polygons between their orbits but was unable to find any geometric pattern which fitted. Then, to his great delight, he found that when he modelled the problem in three dimensions, using the crystal spheres instead of plane circles, he could fit the five regular solids between the spheres:

> And then again it struck me, why have plane figures among three-dimensional orbits? Behold, reader, the invention and whole substance

> of this little book! In memory of the event, I am writing down for you the sentence in the words from that moment of conception: The Earth's orbit is the measure of all things: circumscribe around it a dodecahedron, and the sphere containing this will be Mars; circumscribe around Mars a tetrahedron and the sphere containing this will be Jupiter; circumscribe around Jupiter a cube and the sphere containing this will be Saturn. Now inscribe within the Earth an icosahedron, and the sphere contained in it will be Venus; inscribe within Venus an octahedron, and the sphere contained in it will be Mercury. You now have the reason for the number of planets.[6]

Kepler first met Tycho Brahe early in the seventeenth century, when the latter was living in Prague. The two astronomers only met each other for a short time, but Tycho got to know Kepler well enough to realize that he was the right man to complete and publish the *Rudolphine Tables*, named after his patron, which would enable astronomers to calculate the positions of the planets from his years of observations. When he was on his deathbed, Tycho pleaded with Kepler to finish his work on the planets and to publish after his death. Kepler worked hard at the data, but could not get the observations to fit the cycles and epicycles of Ptolemy. The errors were small, only 6 to 8 minutes of arc, but he knew they were errors and that the circular motions were simply not capable of accurately predicting the positions of the planets. After much trial and error, he tried to fit the data for Mars into an ellipse instead of a circle. To his great joy, the data fitted well and it explained perfectly the error of 6 to 8 minutes of arc.

It was here that Kepler made his great contribution to astronomy, the laws of planetary motion by which he is remembered:

> LAW 1 *The orbits of the planets are ellipses with the Sun at one focus.* The ellipse is simply a projection of a circle but foreshortened in one direction. It has many surprising properties which can only be appreciated from the mathematics. The ellipse has two foci which are symmetrically placed about the centre on its major axis. The circle is a special case of the ellipse where both foci coincide with the centre.

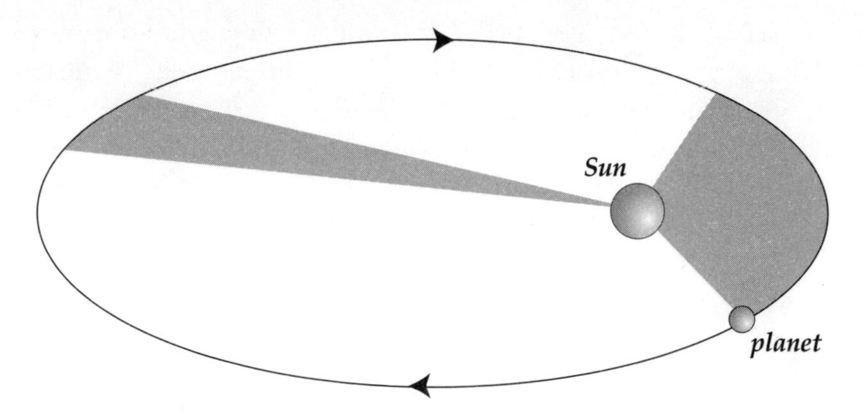

FIG. 2. Kepler's Second Law
Johannes Kepler formulated three laws governing the motion of the planets. The diagram illustrates his second law, that the planet sweeps out equal areas (shaded) in equal times. Thus the planets travel faster when they are nearer to the Sun.

LAW 2 *The radius vector sweeps out equal areas in equal times.*
The radius vector joins the centre of the Sun to the planet. The vector does not rotate at uniform speed but Kepler's law defines its position precisely. It is obvious from the swept area that the planet moves more slowly at greater distances and reaches its maximum speed on its closest approach to the Sun.

LAW 3 *The cubes of the semi-major axes of the planet's orbital ellipses are proportional to the squares of their periods.*
This law is frequently expressed in terms of the distance from the Sun rather than semi-major axes. Whilst the former is easier to understand it is ambiguous because the distance changes throughout the orbit. Kepler was particularly pleased with this mystic numerical proportion.

Kepler produced the most accurate tables ever published for finding the positions of the planets. It might be thought that astronomers flocked to buy his tables, but very few were sold and in the early years few astronomers read his work. By the 1630s there were several rival tables available to find the positions of the planets and nobody had ever compared them to discover which was the best.

In fact, the astronomers who first discovered their merits were the Englishmen Jeremiah Horrocks and his correspondent William Crabtree in the late 1630s. At the time, they were the only astronomers taking accurate measurements of the planetary positions.

There is one other astronomer of importance before we come to Horrocks. He is Galileo Galilei, who was born in Pisa on 15 February 1564. His father, Vincenzo Galilei, was a musician. In the early 1570s the family moved to Florence, where their ancestors had lived for generations. Galileo attended the monastery school at Vallombrosa near Florence, and in 1581 he matriculated at the University of Pisa, where he had been sent to study medicine. His interests did not lie with medicine, however, and, much against the wishes of his father, he decided to make mathematics and philosophy his profession. He prepared himself to teach Aristotelian philosophy, but in 1585 he left the university without obtaining a degree and for several years he gave private lessons in mathematical subjects at Florence and Siena. During this period, he began his studies on the motion of bodies, a topic which he pursued steadily for the next two decades.

Galileo's early career was not exceptional, but in the spring of 1609, when living at Padua, he heard that an instrument had been invented in the Netherlands which was capable of showing distant objects as though they were nearby. By trial and error, he quickly deduced the principle of the telescope and he made his own spyglass from lenses that he bought in a spectacle-maker's shop. Other experimenters had done the same thing, but Galileo worked out how to improve the instrument. At this time, he was working at the university in Venice. He taught himself the art of lens grinding and went on to produce a range of telescopes which were increasingly more powerful.

By August 1609 he had created a telescope with a magnification of eight, which he demonstrated to the Venetian Senate. The merchants of Venice could use Galileo's telescope to sight their ships on the horizon and to gain a commercial advantage by identifying the ships before their rivals. They were so pleased with the instrument that Galileo was rewarded with a life tenure and a doubling of his salary. It made him one of the highest paid professors at the university. Galileo continued to improve his telescopes and soon he was observing the heavens with instruments which magnified up to twenty times. He found that the telescope showed many more stars than are visible with the naked eye. He drew the Moon's phases as seen through the telescope, and he could tell from the shadows of the crater rims on the Moon that its surface was not smooth, as had always been supposed, but rough and uneven. He trained his telescope on Jupiter and discovered four moons revolving around it. Through the telescope, he could see that Venus had phases similar to the Moon, and this was one of the main factors which convinced him that the planet Venus orbited the Sun. Galileo produced a little book called *Sidereus Nuncius* (The Sidereal Messenger), in which he described his findings. He dedicated the book to Cosimo II de' Medici (1590–1621), grand duke of Tuscany, whom he had tutored in mathematics for several summers, and he further honoured the Medici family by naming the moons of Jupiter after them. Galileo was rewarded for his efforts with an appointment as mathematician and philosopher to the grand duke, and in the autumn of 1610 he returned in triumph to his native state. He became a courtier and was able to live the life of a gentleman.

Galileo's discoveries did not prove directly that the Earth was a planet orbiting the Sun, but they did a great deal to undermine the cosmology of Aristotle. It was obvious to him that the phases of Venus, clearly seen through his telescope, proved that the planet orbited the Sun. Galileo became convinced in his belief that the Sun was the centre of the universe and that the Earth was a just another planet, as Copernicus had argued. Galileo's conversion to Copernicanism was a key turning-point in the scientific revolution. The pope gave Galileo permission to write a book about his theories of the universe, but warned him to treat the Copernican theory as no more than a hypothesis. The book, *Dialogo sopra i due massimi sistemi del mondo, Ptolemaico e Copernicano* (Dialogue Concerning the Two Chief World Systems, Ptolemaic & Copernican), was finished in 1630 and published in 1632. It was cleverly presented as a dialogue between Aristotle, Ptolemy and Copernicus, so that the views expressed were not necessarily those of Galileo. The presentation did not please the authorities at the Vatican, however. It was obvious that Galileo was claiming that the Earth moved, a teaching which was contrary to the account of the fixed Earth given in the Bible. It generated the famous rift between Galileo and Pope Urban VIII.

At the age of seventy, Galileo was put under house arrest. He was forced to withdraw his teachings and to sign a document to say that they were untrue. He knew perfectly well that his theories were right, but the intervention of the Roman Church was one of the main reasons why the next advances in astronomy appeared in the Protestant countries of northern Europe as opposed to the Catholic countries of the Mediterranean.

CHAPTER FIVE

Horrocks Studies the Heavens

During his time at Cambridge, Jeremiah Horrocks came across a treatise written by D. H. Gellibrand, the professor of astronomy at Gresham College in London. In his treatise, Gellibrand praised the works of a Belgian astronomer called Philip Lansberg. Horrocks wrote to Gellibrand, who recommended Lansberg's tables to him as the most up-to-date work. With some difficulty, Horrocks managed to obtain a copy of Lansberg's work, *Progymnasmata de motu Solis*, and another book by Lansberg, *Uranometriam, Tabulas Perpetuas.*

Horrocks's copy of Lansberg's tables has somehow managed to survive and find its way into the library of Trinity College, Cambridge (see list on p. 63). Horrocks purchased it in 1635, at the end of his final year at Cambridge. Inside the cover he listed thirty-one books. It seems improbable that he could have acquired so many volumes so early in his career, and indeed he did not own many of them at all. A handwritten footnote explains that the list is a catalogue of books which he knew to exist and which he wanted to read. The list was doubtless compiled from the Cambridge library catalogues and from the bibliography in Lansberg. Many of the books dated back to the previous century.

When the list is compared to the library at Emmanuel College, with its straining and heavily overloaded theological shelves, it is obvious that Horrocks must have managed to consult astronomical books in other Cambridge libraries. None of the books on his list were published in English and nearly all the titles are in Latin. This underlines the advantage of Latin as an international language for communication in the seventeenth century. Very few academic works were written in the language of the country in which they were published.

Many of the titles are very obscure after the lapse of nearly four centuries, but the names of Ptolemy and Copernicus are easily recognized. There are two works by Tycho Brahe and four by Johannes Kepler, including his *Optics* and also the *Rudolphine Tables.* Among the lesser known astronomers are Longomontanus, who worked on the observational data of Tycho Brahe, and Lansberg, the author of the volume in which the list was written. The reason why very few modern astronomers have heard of Philip Lansberg will soon become apparent. It is of interest that not all the works in Horrocks's list are on astronomy. He seems to have had a taste for the ancient Greek historian Herodotus and for Pliny's work on natural history. Galileo's work, his *Dialogue Concerning the Two Chief World Systems,* published in 1632, is conspicuous by its absence, though this work may not have reached England at the time Horrocks compiled his list.

It is clear that Horrocks had found references to Kepler during the three years he was at Cambridge, but neither he nor anybody else at that time realized the significance of Kepler's work on planetary orbits. Even Copernicus's *Revolutionibus,* which had been in print for ninety years, was not well known in England outside a small circle of astronomers. The plain fact was that in Horrocks's time the hard-won knowledge of the great astronomers, which we now accept so readily, circulated painfully slowly. In the early 1600s the great majority of people in England and the rest of Europe, even the better educated, still thought that the Earth was at the centre of the universe. The revolutionary theories had reached Cambridge, but they were still too radical to be accepted by the establishment. Jeremiah Horrocks, however, seemed to have an instinctive feeling for the new astronomy and right from the outset of his investigations he assumed that the Copernican system was correct. This may not seem very significant, but we must bear in mind that academics in Italy, with the exception of Galileo, did not believe that the Sun lay at the centre of the universe. It was an idea which had been rejected even by men as forward-thinking as Bacon. It is a sobering fact that a very young

Horrocks's Astronomical Catalogue, 1635

1 Albategnius [Arabian astronomer, fl. AD 880] [published 1537]
2 Alfraganus [Arabian astronomer, fl. AD 800] [published 1592]
3 J. Capitolinus [Roman biographer]
4 Clavii [German mathematician]: *Apologia Calendarii Romani* [published 1588]
5 Clavii: *Comm. In Sacroboscum* [abridgement of Ptolemy, published at Cologne in 1601]
6 Copernici [Copernicus]: *De Revolutionibus* [published 1566]
7 Cleomedes [Greek geometer]
8 Julius Firmicus [astrologer of the fourth century]
9 Gassendi: *Exerc. Epist in Phil. Fluddanam* [Gassendi was the first astronomer to observe a transit of Mercury]
10 Gemmae: *Friscii Radius Astronomicus* [a work on astronomical Instruments]
11 Corneli [Flemish astronomer, 1535–77]: *Gemmae Cosmocritice*
12 Herodoti [Herodotus, the classical Greek historian]: *Historia*
13 J Kepler: *Astron. Optica.* [Kepler on optics]
14 J Kepler: *Epit. Astron. Copern.* [Kepler on Copernicus]
15 J Kepler: *Com. De motu Martis* [Kepler on the motion of Mars]
16 J Kepler: *Tabulae Rudolphinae* [Kepler's *Rudolphine Tables*]
17 Lansbergii: *Progymn. de motu solis* [Tables of Lansberg]
18 Longomontani: *Astron. Danica* [Tables of Longomontanus]
19 Magini [Venetian astronomer]: *Secunda Mobilia* [published 1585]
20 *Mercatoris Chronologia* [Mercator, of projection fame, published 1569]
21 Plinii: *Hist. Naturalis* [Pliny's *Natural History*]
22 Ptolemaei: *Magnum Opus* [Ptolemy, *Almagest* or 'Great Work']
23 Regiomontani: *Epitome* [Ptolomaic system by John Muller (1436–76)]
24 Regiomontani: *Torquetum* [a second volume by Muller]
25 Regiomontani: *Observata* [Muller's observations]
26 Rheinoldi [Rheinold of Saxony]: *Tab. Prutenicae* [planetary tables]
27 Rheinoldi: *Commentaria in Theor. Purbachii* [Astronomical Theory, published 1585]
28 Theonis Comm [Alexandrian astronomer]: *In Ptolom.* [Ptolemaic system]
29 Tycho Brahea: *Progymnasmata* [Works of Tycho Brahe]
30 Tycho Brahea: *Epist. Astron.* [Brahe's Astronomical correspondence]
31 Waltheri: *Observata* [Observations by a pupil of Muller published 1618][1]

provincial astronomer in England embraced the new philosophies of the universe before they were accepted by the university professors of Florence and Venice.

In 1635, at the age of seventeen, Jeremiah Horrocks returned to his native Lancashire. All he wanted from his life was to be able to study the stars and planets. He needed his books on astronomy to achieve his aims and he also needed instruments. The latest thing in astronomical instruments was the telescope. It had been in use for about twenty years and it was beginning to become far more readily available. He tells us that he purchased a 'half crown' telescope, probably at a local fair. This was at a time when the wages of a farm labourer amounted to only a few shillings a week. The telescope was therefore a modestly priced item compared to the cost of the books he had to buy. The popular fairground telescope was designed as little more than a toy with which to view things at a distance, rather than an instrument with which to study the stars. It seems that early in his career Horrocks purchased this cheap fairground telescope and, finding the instrument to be very valuable for his purposes, went on to procure the best instrument he could find. In May 1638 Horrocks wrote that 'I have at last obtained a more accurate telescope' and a year later he tells us that 'the telescope which I employed on this occasion is much more accurate than those generally used'.

Other astronomical instruments were much more difficult to come by. Liverpool was a seafaring town, so the astrolabe and cross staff were probably on sale at the local fairs and markets, and even if this was not the case, there were plenty of merchants and sea captains who could easily procure these navigational instruments for him. He may have already purchased these items at Stourbridge Fair during his time at Cambridge. The problem facing Jeremiah Horrocks was how to obtain more sophisticated astronomical instruments, for which, in the 1630s, there was no market at all. His only option was to make his own.

In the second half of the century, it was the makers of the familiar English long-case clocks who made specialized astronomical instruments. Jeremiah could not have been more fortunate. His father was a watchmaker, as were his uncles. He was surrounded by watchmakers on both sides of the family and they were amongst the leading practitioners in the country. They had specialized tools and professional expertise. The making of astronomical instruments was a welcome and interesting diversion for them. It was a simple task after the intricate and minute detail of the watches, but one which still demanded the great accuracy to which they were accustomed to work. Jeremiah Horrocks was therefore free to design his own instruments, to discuss the design with his father, who would construct the instruments for him, or possibly involve one of his uncles in the manufacture. Perhaps he made them under supervision with his own hands, using the specialized tools of the watchmaker. The family saw a young man who had returned from college with an obsession about astronomy, and the evidence shows that they were prepared to support him in his chosen calling. It is reasonable to deduce that Jeremiah had to work in return for his keep – he would have helped with the family business, with the repair and calibration of the watches and their marketing – but working for his father gave him much more freedom than any other form of employment. His hobby of watching the stars was a night-time vigil, in the darkness, when it was impossible to work on the watches. His daytime job was working on the watches, when the light was good but it was impossible to see the stars.

Horrocks owned a radius astronomicus. This was a simple instrument, a development of the Elizabethan cross staff, consisting of no more than a wooden cross or, in Horrocks's case, a 'T' shape, with a sighting at the end of the long arm and with two movable sights on the cross piece. The cross piece was straight, not a circular arc as the name suggests. It had a graduated scale and could be used to measure the angle between two bearings or two stars. Horrocks mentions that his radius astronomicus was three feet in length and the scale was divided into ten thousand equal parts. A simple calculation shows that the distance between the divisions was a tenth of a millimetre and a further calculation shows that in theory he could measure angles down to a fraction of a minute of arc. This was an impossibly small angle which could never be achieved in practice, but he came fairly close to the limits of the instrument and he was soon measuring angles within one or two minutes of arc. He was obviously pleased with his radius astronomicus, for in January 1637 he mentioned that he had made a larger instrument eleven feet in length.

His first task was to measure the positions of the planets against the background of the fixed stars. The stars had been carefully plotted and recorded by Tycho Brahe, and Horrocks's aim was to measure the angle between the planet's position and two or three neighbouring stars. We now know that the stars do have very small proper motions in the sky, but their positions had not changed measurably since the time of Tycho Brahe. To all intents and purposes, the stars were fixed and unchangeable, Horrocks was therefore able to use the stars as a set of fixed points from which he could plot the motion of all the visible planets across the sky. He had been reading a book on navigation by Edward Wright, who pointed out that small errors can occur through the position of the eye. This was called the parallax of the eye. Horrocks noticed that there was a difference of 7 minutes of arc when he used the left side of his radius astronomicus compared to the right side. He tried to improve his measurements by allowing for this parallax:

> Venus was 10° 29' distant from Saturn. I applied the radius astronomicus to my eye so that the surface of my eye was at the end of the radius, as I usually observed. But while I observed there came to my mind what Kepler says on the eccentricity of the eye and what our Edward Wright has in his book on errors of navigation. That I might avoid this mistake I immediately did this: I set up a pin at the very end of the radius and, keeping the radius in exactly the same position (which was not difficult by reason of the convenient position) I moved my eye backwards and forwards until I saw the pin in line with both sights and the star: This done I found the distance was 10° 22' as accurately as I was able to estimate on account of the bright rays from Venus. So likewise if I moved the end of the radius to the right hand corner of the right eye so that the end of the radius was behind the surface of the eye, the distance I thus found was 10° 15'. Thus eccentricity of my eye caused an error of 7' in my radius set at 10 ½°. And consequently if the three foot length is divided

> into 10,000 parts (as is my radius) the rays will be seen to meet at 102 parts beyond the end of the Radius[2]

He had other measuring techniques, but the radius astronomicus was his preferred instrument for measuring angles. He wanted to measure the diameters of the Sun and the Moon as accurately as possible and he devised a system of plumb lines with vertical silk threads touching the edge of the Sun and the Moon. The Moon could be viewed directly with the naked eye, but to measure the Sun he used a foramen technique, which involved projecting an image of the Sun through a small hole and onto a screen – effectively a pinhole camera. He did not persevere long with this method, for he soon discovered that when he wanted to project a bright image on to a screen, his telescope lenses gave him far better results. He discovered that there was a small cyclic variation in the diameters – more noticeable in the case of the Moon – but he arrived at a mean diameter of 31 minutes of arc in both cases.

Jeremiah Horrocks soon ran into deep trouble with his observations of the planets. The basic problem was he encountered the same errors as Kepler had met when the latter was trying to fit the Ptolemaic system of cycles and epicycles to Tycho's observed positions. The planets were simply not in the positions which Lansberg's tables predicted. Horrocks repeated his measurements and his calculations. The errors were only of the order of a few minutes of arc, but Horrocks, like Kepler before him, knew that the tables were wrong and he was very concerned about the discrepancy.

After working on his own for nearly a year, Horrocks was introduced to William Crabtree, a man about eight years his senior. Crabtree was a draper who lived at Broughton, near Manchester. He was born in 1610 and it is probable that he was educated at Manchester Grammar School, though the school records do not confirm this. He did not go to university but in his times a university education was only of indirect help to the astronomer. He was a skilful observer and he could handle the necessary mathematics. When he and Horrocks were introduced, Crabtree had been married to Elizabeth Pendleton for three years. Some accounts say that it was the scientist

Astronomicall Exercises. 1

1

P. Lansberg sayth[1] ÿe fixed starres do move equally every day 8'''' 28'''' 12v 32vi. and performe one Revolution in 28000 yeares. But these two cannot stand togethere: for according to the dayly motion he appoints and useth in all his Observations of ÿe fixed starres, one Revolution will be made in

Sexagens of days 42''' 45'' 16' 32d. wch is
Julian yeares 25284 days 11.
Ægyptian yeares 25301 days 127.

In which time ÿe Sun moves in his equall motion Sexagen. 42''' 8'' 27' 22° 4'.
That is, 25284⅘ Revolutions. There is therefore some mistake herein, that is the cause of this error and contradiction: for if he had sayd 28000 rotundo numero, he should rather have set down 25000. Learne hence how unsafe it is to take any thing upon trust, though from most learned men

2

The distance of the fixed starres from ÿe Earth (or rather from ÿe Sun) Lansberg[2] maketh to be 28000000 such parts, as the Semidiameter of the Earths great orbe (or ÿe mean distance of ÿe Earth from the Sun) is 10000. But taking the abovesayd proportion of the Period of Revolution, between the Earth and the fixed starres, ÿt is to say 25284⅘ to 1. and supposing as Lansberg doth[3], that this proportion is answerable to the Semidiameter of their orbs, I say then their distance from the Sunne will be 252845556 feere. such parts as the Radius of the Earths great orbe is 10000: But this being 1498½ (ÿe Suns[4] mean distance from ÿe Earth according to Lansberg) the other will then bee 37888906½. which is the distance of the fixed stars from ÿe Sun in Semidiameters of ÿe Earth.

[1] Thes. Obs. Astr. p. 160.

Stellæ fixæ.

	/	o	/	//	///
42'''	5	53	38	46	24
45''		6	18	54	24
16'			2	14	43
32d				4	29
	6	00	00	00	00

Sol

	///	//	/	o	/
42'''	41	23	49	49	29
45''		44	21	14	49
16'			15	46	13
32d				31	33
	42	8	27	22	4

[2] Uranomet. p. 124. l. 3

[3] Uranom. p. 125. l. 7

[4] Uranom. p. 58.

FIG. 3. Sheet from 'Astronomical Exercises'
Page from Horrocks's notebooks concerning the apparent rotation of the stars and Lansberg's tables.

and antiquarian Christopher Towneley who introduced Horrocks to Crabtree, but this is improbable since Horrocks was completely unknown at the time. There is a much more likely possibility. John Worthington, whom we met as a close friend of Horrocks at Cambridge, was the son of a Manchester cloth merchant. Seventeenth-century Manchester was a town no larger than Liverpool, and it is certain that Worthington senior and William Crabtree knew each other. They were in the same line of business, and it therefore seems very likely that the introduction came through that channel. It was William Crabtree who made the first approach, but his letter to Jeremiah Horrocks does not survive. Horrocks's reply does survive, however, and is dated 21 June 1636. It shows that Horrocks was feeling the isolation of working on his own and was on the point of giving up astronomy altogether. The introduction to Crabtree was very important to him. He badly needed somebody with whom he could try out his ideas:

> I received your letter dating from the beginning of this month. Although you may think it needs some excuse to justify it, in my judgement it deserves praise instead. As someone who wants to be numbered amongst the ranks of practitioners of astronomy, I would be unworthy indeed if I spurned anyone else engaging in the same field of study. Certainly, I have been engaging for some time in this activity and have really been doing so only to gain experience and give myself pleasure, and it has seemed to me that there was no delight missing in this agreeable activity other than that I lacked some colleague who might join forces with me and engage in the same work. As (unhappily for me) I had not yet found such a person (although one has been much sought after by me), this might have been enough for me to lose heart and, with no colleague in support, to cease applying my energies any further in my endeavours. And indeed I was already beginning to give up my studies or at least to pursue them in a less committed way, when your unexpected support put new heart into me so that I could take up work once again. So, I warmly welcome both the friendship that you offer and your own labours and I am full of happiness, and I promise that I will give you (or any others too, if such

> there be, who would wish to devote their energies in this direction) as much help and assistance as I can.[3]

Horrocks and Crabtree started to exchange regular correspondence and were soon on excellent terms with each other. It was nearly two years before they met face to face, in June 1638. Jeremiah Horrocks communicated his findings and his dismay regarding Lansberg's tables to his new friend. It happened that Crabtree had encountered exactly the same problem regarding the positions of the planets, but he was ahead of Horrocks and suggested that instead of using Lansberg's tables Jeremiah should use the *Rudolphine Tables* published by Johannes Kepler. Crabtree was an accurate observer and a good mathematician, but he was an amateur astronomer. Yet here we find a provincial cloth merchant who was more up to date on the state of the art than the professor of astronomy at Gresham College. Gellibrand, the professional, had done no more than anybody else would have done in his position. He had recommended the latest publications to Horrocks, but his duties at Gresham College did not allow time for him to do any actual observing and to check them out for himself. The only way to judge the best of the several systems available was to actually measure the positions of the planets in their orbits and to compare the data with the predicted values. In the 1630s Horrocks and Crabtree were the only people in the country – in fact, the only astronomers in the world – taking the necessary measurements. This was how they came to be the first astronomers in England to appreciate the work of Kepler. As Horrocks and Crabtree investigated Kepler's system further, they were delighted with their findings. They found that the planets followed the elliptical orbits with great accuracy, exactly as predicted by Kepler in the *Rudolphine Tables.*

Horrocks knew all about the Earth-centred world and the cycles and epicycles envisaged by Ptolemy. The geocentric world of the ancients was much smaller and required far less space than the heliocentric universe of Copernicus. But after the months he had wasted on Lansberg, Horrocks was convinced from his observations that Ptolemy's system was wrong. Jeremiah Horrocks became very cynical about what he unfairly called the 'pagan' Ptolemaic system. 'Having long contemplated and admired a philosophy so sublime and so worthy of a Christian I thus expressed my aversion to the puerile fictions of the pagan Ptolemy.'[4] We now discover a new and unexpected dimension in the personality of Jeremiah Horrocks. He had a talent which is very rare in an astronomer. He proceeded to express his feelings in blank verse:

> Why should'st thou try, O Ptolemy, to pass
> Thy narrow-bounded world for aught divine?
> Why should thy poor machine presume to claim
> A noble maker? Can a narrow space
> Call for eternal hands? Will thy mansion
> Suit great Jove? or can he from such a seat
> Prepare his lightnings for the trembling Earth?
> Fair are the gods you frame forsooth! nor vain
> Would be their fears if giant hands assailed them.
> Such little world were well the infant sport
> Of Jove in darker times; such toys in truth
> His cradle might befit, nor would the work
> In after years have e're been perfected,
> When harlot smiles restrained his riper powers.

Where on earth did Horrocks learn to express himself in such fine language? Was it from the sermons of the great Puritan orators Richard Mather and John Cotton? If so, then perhaps some of the long-forgotten sermons of the seventeenth-century Bible-thumpers are worthy of study after all. But surely we can hear echoes of Shakespeare in the verse. The bard died two years before Jeremiah was born, but since the publication of the First Folio he had become very popular and there were many alive who still remembered him. These lines are only a starter. Horrocks was by no means finished with his criticism of Ptolemy:

> These are your fancied gods, your paltry dreams;
> And worthy them is all you raise around;
> The temples that you build are amply large,

Thy heavens are suited to a Jove like thine.
Are such the auspices by which you rule
Your world? No longer I deplore the Earth
That stands begirt with solid adamant;
Such walls repel unholy deities,
And keep the nations pure. How wisely doth it
Court repose far from the stars where it would
Have to mingle in degrading commerce,
And find not heaven, but realms replete with crime,
Calm urge thy chariot through the starry sphere

The original lines were written in Latin. The translation given here is the work of Horrocks's Victorian biographer, the Revd Arundell Blount Whatton. These are not, therefore, the exact original words of Jeremiah Horrocks which appear in the poetry, but it is obvious that Whatton has done a very good job on the English translation. Horrocks ends by reflecting on the task of keeping the stars and the planets in motion:

No trifling task it is to hurl at once
So many gods and stars in uniform
Gyration. Then let those whose little sum
Of learning reaches but to tell the tale
Their fathers told before, whose every word
Deals in absurdities unworthy heaven,
Rival each other to applaud this fable.

In a beautiful piece of blank verse, he breaks into ecstasies about the 'projected Earth rolling along its planetary path', followed by an oblique reference to the telescope. In an accurate descriptive passage, one of his best lines is the way in which he explains that the spinning of the Earth on its axis 'saves to all the distant stars the useless labour of unceasing motion'. Horrocks allows himself to get carried away as he describes the beauty of the Copernican system:

But a sublimer throne is thine, and awe
Ineffable awaits thy lightning's course,
Thou God of truth, whose certain laws direct

The starry spheres, whilst all the powers above
Admire and tremble; the projected Earth
Rolling along its planetary path
Hath learned to hail thy triumph; and this age
Enables mortal eyes in thy great works
To view thee nearer, and with nobler thought
To trace the stars whose order proves them thine.

In vain the Sun his fiery steeds would urge,
In vain restrain them, or attempt to guide
Their rapid course within the laws of fate.
The Earth performs their task, and by each day's
Revolving saves to all the distant stars
The useless labor of unceasing motion.

The clouds which once obscured our mental sight
Are gone for ever; great Copernicus,
Sent from above, lays open to our view
The arduous secrets of wide heaven's domain
Turn hither then your grateful steps, for here
Are wondrous mysteries that you may learn,
Open to all whom, freed from baser thoughts
The love of truth impels, and whom no cry
Of vulgar men can scare from what is right
Nor fear oppress, O child, of ignorance!
Nor fabling oracles once deemed divine.[5]

It was the seventeenth century. Any astronomer worth his salt also had to be a philosopher. It was all very well trying to explain the motions of the planets and the system of the world, but any model of the universe had to explain how and when the Sun and the planets had been created in the first place. Kepler had tried to explain the Creation in his writings and so, too, had a number of leading theologians. Horrocks also wanted to give an account of the Creation, and his work has survived in a notebook which he called his 'Philosophical Exercises'. His account is very brief, but it enables us to see something of the working of his mind. It is interesting historically as it expresses contemporary ideas about the Creation.

It could be argued that there was no need to develop a new theory of the Creation. The book of Genesis explained everything about the beginnings of the world and any theory which did not agree with the Bible was by definition heretical. This was the very time at which Galileo was threatened with excommunication for teaching that the Earth orbited the Sun. The account in the first chapter of Genesis states that 'God made two great lights: the greater light to rule the day, and the lesser light to rule the night: he made the stars also.' The Earth was created on the first day, the Sun and the Moon, according to the biblical account, were created three days later, on the fourth day. This implies that they were lesser bodies than the Earth. Their function was to provide light in the sky, the Sun to rule the day and Moon the night, but their size is left as a matter of interpretation: the Bible does not state it explicitly. Jeremiah Horrocks had no doubt that the Copernican theory of the universe did not conflict with the Bible, but in any case the last thing the Puritans wanted was to have their views dictated by Rome. Any suggestion that Horrocks might be guilty of heresy simply did not arise: the Puritans did not see it as part of their creed to debate the details of the Creation, and in any case, his work was unknown and obscure. He discussed his ideas with his friends and family and he wrote them up in his notebooks, but nobody had the opportunity to read his thoughts and he was a long way from getting them published.

One of the leading theological thinkers of the times was James Ussher, the archbishop of Armagh, who became primate of all Ireland in 1634. Ussher studied the genealogies in the Old Testament. The generations were given right up to the birth of Christ and these could be used to estimate the time which had elapsed since the Creation. It was easy to work out a chronology for the first generations after Adam, but in the later generations the ages of the parents were seldom given and most of the chronology was therefore reduced to rough estimates. This was only a minor deterrent to James Ussher and he estimated that a total of 4,004 years had elapsed from the time of the Creation to the birth of Christ. The Creation therefore took place in the year 4004 BC. In fact, he went further than this and specified the date as 22 October and the instant of the Creation as sometime in the evening of that day.

James Ussher was born in 1581. He was thirty-seven years older than Jeremiah Horrocks and it might be reasonable to assume that Horrocks knew of his work and that Ussher was the inspiration behind Horrocks's ideas. Ussher's chronology was not published until the 1640s, however, after the death of Jeremiah Horrocks. Thus, the astronomer's ideas were not directly influenced by Ussher, and Horrocks makes no mention of him in his 'Philosophical Exercises'. There are so many similarities between the thinking of the astronomer and the theologian, however, that it can only be assumed that their ideas came from the same source – the current ideas being put around at the time. Johannes Kepler also gave an account of the Creation of the world, but he disagreed with Ussher on one point. He claimed that the Creation took place in the summer and not the autumn. Horrocks had studied Kepler's account carefully. He agreed with most of it and he accepted that the time scale of the world spanned about six thousand years from the Creation to the Day of Judgement, but he was convinced that the world was created in the spring and he took issue over some of the points raised by Kepler.

> To this I answer, with Kepler, that every thing naturally desires to rest in ye same place where ye hand of nature hath at first placed it. Thus ye Sun being placed in ye centre of ye world doth abide there unmoved, because there is no creative power wch is able to remove it out of its place, it being ye greatest & strongest body in ye world. But ye stars though they have a proper place assigned to every one of them, wch they do, in imitation of ye Sun, labour to keep, & having lost, to recover, yet being but little bodys, ye Sun easily prevails over them and carrys them about at its pleasure yet so as ye planets do in two things retain some prints and footsteps of their captivated liberty, that is there slownes in motion, proportioned to their distance from ye Sun, & there labouring to keep their station assigned by nature, wch is ye cause of their eccentricitys.[6]

When he came to study the motion of the stars, Jeremiah Horrocks thought that they were very small bodies and he had no doubt that they orbited around the Sun. They appeared to take 26,000 years to

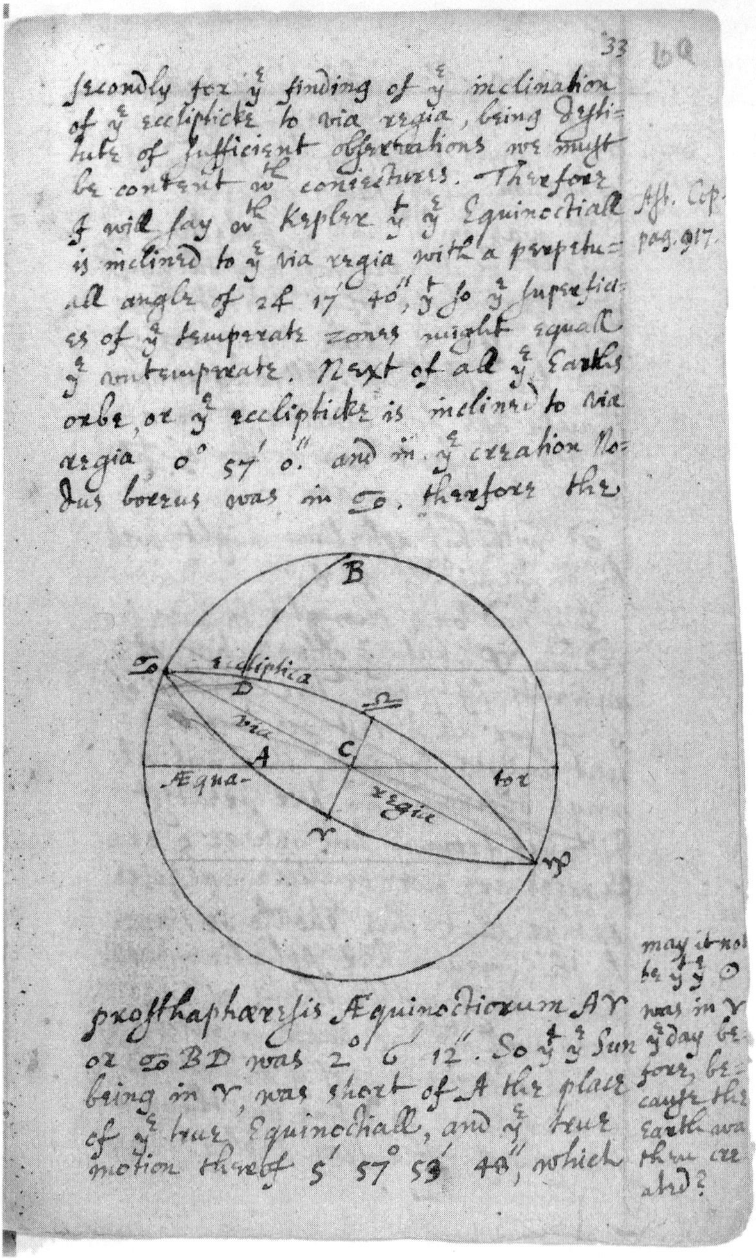

33

secondly for ye finding of ye inclination of ye ecclipticke to via regia, being destitute of sufficient observations we must be content wth conjectures. Therefore I will say wth Kepler yt ye Equinoctiall is inclined to ye via regia with a perpetuall angle of 24 17′ 40″, yt so ye superficies of ye temperate zones might equall ye untemperate. Next of all ye Earths orbe, or ye ecclipticke is inclined to via regia 0° 57′ 0″ and in ye creation Nodus boreus was in ♋. therefore the

Ast. Cop. pag. 917.

prosthaphaeresis Æquinoctiorum A♈ or ♋BD was 2° 6′ 12″. So yt ye Sun being in ♈, was short of A the place of ye true Equinoctiall, and ye true motion thereof 5′ 57° 53′ 40″, which

may it not be yt ye ☉ was in ♈ ye day before, because the Earth was then created?

FIG. 4. Sheet from 'Philosophical Exercises'
Page from Horrocks's notebooks building on the work of Kepler to try to calculate the moment of the Creation.

complete their orbits, an illusion which is actually the result of the precession of the Earth's axis of rotation. This very long period was seen as an indication of the great distance to the stars, and he used this model to make an estimate of how far away they were. When he came to consider the creation of the planets, he put forward the theory that every planet was formed in its aphelium – that is, at the point in its orbit where it is at the greatest distance from the Sun. This, he argued, was the reason why the planets moved more slowly at greater distances from the Sun: they had a natural desire to linger as long as possible at the place where they were born. He gave five reasons why he thought Kepler's account of the Creation was wrong:

> 1. His was ye first day of ye weeke, but ye stars were made ye fourth. [Obviously a reference to Genesis]
>
> 2. The motions of ye planets will not agree unto that beginning, and the observations of ye antients doth, and I thinke, we can hardly make so bold with ye antients as to bring the planets motions precisely to quadrants especially in Mars.
>
> 3. It is more probable that ye true motions rather then ye equall should have been in quadrants precisely, if God had respected that concinnity.
>
> 4. The magnitudes and distances of ye Sun Moon & Earth, Keplers speculations deduce from ye Sun and Moons apparent semidiameters in their apogaeum, & yet in ye creation, when ye eye of nature, if ever, should have regarded that pulchertude, they are both made in their mean distance: can nature in it's working, have respect to those Archetypes, which are not, perhaps never will be precisely.
>
> 5. The world is most likely to be created in ye spring, God commanding ye Isrealites at that time to begin their yeare. But his [Kepler's] was in ye summer.[7]

Horrocks found it confusing to deal with dates before the birth of Christ, so he introduced a new time-scale of 6,000 years. He went on to claim that the moment when the stars were created was 'in ye beginning of ye 4th day of the week, which I make midnight preceding, and that in ye meridian of Babylon, where it is thought, Paradise stood'.[8] He made the date 23 April in his new time-scale, in the year of the Great Period 1944 when it was midnight in the meridian of Babylon. He does not, unfortunately, tell us the date of the birth of Christ in his Great Period, so we cannot compare his time-scale with that of Ussher. Nor does he make it clear what happened in the 1,944 years prior to the Creation.

It is easy to criticize the naïvety of Horrocks's ideas about the age of the Earth and the Creation, but his work must be viewed against the context and ideas of his times. It was another two centuries before scientists began to realize the incredible time-scale required for life to evolve on Earth, the immense age of the rocks and the time it took the stars and planets to form.

Horrocks thought that he could measure the maximum and minimum distances of the planets from the Sun and that these limits would be of help to astrologers. It was almost impossible in Horrocks's time to differentiate between astronomers and astrologers. The astrologers wanted to determine the positions of the planets as accurately as possible so that they could make more accurate predictions when casting their horoscopes. Early in his career, Horrocks showed little sympathy towards the astrologers. He knew he could make no impact on their beliefs and he dismissed what he called the 'contrary faction' and their 'boysterous fopperys':

> I have purposed to combine in a league against those of the contrary faction, that so we might be more naturally have judged what planets had been frends to mans nature, (which I intrusted the Earth, our planet, withall) and which of them were averse: And before this tribunall, Saturn & Mars should have tryed to acquitte themselves of that imputation of malice and evill nature, Jupiter and Venus to have justified their claim to loving and kind effects, and Mercury should have pleaded for his eloquence . . . it appears: that Saturn ate his own children, Mars was bloudy & cruell, & therefore astrologers can, without making syllogisess, prove that Saturn and Mars are no great frends to humane nature . . . it were against common sense to imagine Jupiter and Venus any other then especiall favourers of mankind, especially seeing they are fayr bright stars too; and indeed if Mercury was ye God of eloquence, it were a

miracle if Mercury were tongtied ... I had thought to have refined these boysterious fopperys, but I despair of effecting it, and therfore Astrologers may sleep on, I will not trouble their dreames, for it is some vexation to be awakened out of a pleasing one. Well I awake myself.[9]

The progress made by Jeremiah Horrocks was remarkable, especially given that he was still in his teens, but much of this progress was in the wrong direction. We know that by 1636 he had studied Lansberg, had mastered the *Rudolphine Tables* and was studying Kepler. He had spent a valuable year of his life trying to fit his observations to tables published by people whom he came to see as boastful and incompetent astronomers and he had met with nothing but frustration. At the end of 1637 he wrote a Latin treatise against Lansberg's tables with the title *Jeremiae Horroccii Anti-Lansbergianus, sive disputationes in Astronomiam.* He changed his mind about the content and remodelled it under the title *Astronomiae Lansbergianae censura et cum Kepleriana comparitio.* He rewrote it a second time, introducing the diagram of Hipparchus, and gave it yet another name, but it essentially consisted of the same arguments exposing Lansberg's errors.

He gave credit to Lansberg for explaining the Diagram of Hipparchus, from which Lansberg claimed it was possible to use a

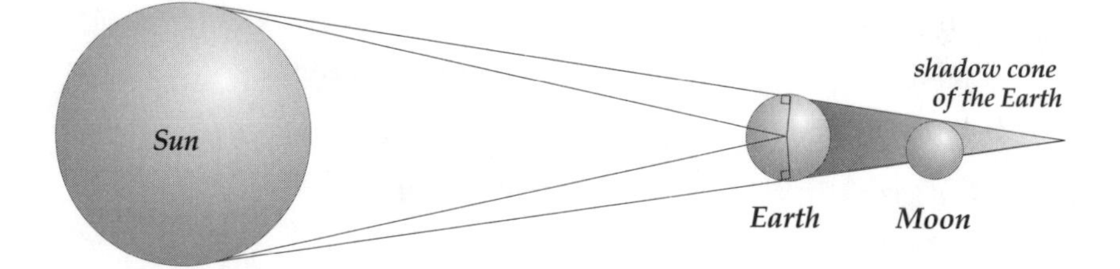

FIG. 5. Hipparchus's method of calculating the distance to the Sun
The method was to measure the angle of the Sun seen from the Earth and also as seen from the apex of the Earth's shadow cone. The tiny difference between these angles enables the distance to be calculated. The shadow cone has to be measured by observing the radius of the Earth's shadow on the Moon, a very difficult observation. Astronomers did not know that the angular difference they sought was only a few seconds of arc.

lunar eclipse to measure the distance to the Sun. Other astronomers, such as Albategnius, Copernicus, Tycho Brahe and Longomontanus were not able to explain the diagram and they disagreed amongst themselves as to the correct interpretation of Hipparchus' work. The controversy surrounded the angles, the correct size of the Sun's disc and the angle of the cone of the Earth's shadow falling on the Moon. But Horrocks was able to show that even if the angles were measured correctly and with sufficient accuracy, the whole method was at fault. In his 'Astronomical Exercises', he also criticizes the figures given by Lansberg in his star catalogue. It is astonishing how somebody so young could go through a learned thesis by a leading astronomer and pick out the minor errors with uncanny accuracy.

> He makes the distance of Occidentalion pleiadum from prima Arietis to be 25° 54', and yet the true place therof is Aries 29° 49', and Prima Arietis is in Aries 4° 25', the difference of which is 25° 24'. There is therfore some mistake eyther in Lansberg or the printer; and instead of Aries = 9° 49', we must read Taurus 0° 19' for the distance from prima Arietis 25° 54' is true, as may appeare in the fourth observation of ye Moons appulse to ye pleyades: And likewise it would not agree with the places of the other pleyades if it were not in Taurus 0° 19' as may appeare plainly by the ensuing scheme. He makes the distance of Australion Pleiadum from Prima Arietis 26° 18', and yet his place is but in Taurus 0° 39'. Here again is another mistake: we must read Taurus 0° 43', for so it will agree with the places of the other pleydes, wch we may thus amend.[10]

He also read a treatise by an astronomer called Martin Hortensius, who criticized the work of Tycho Brahe. Horrocks's opinion of Hortensius was even less than his opinion of Lansberg and he wrote a treatise supporting Tycho against Hortensius, *Pars prima in qua respondítur Maertinii Hortensii cavillis adversus Tychonem.* As with the papers on Lansberg, he produced several drafts of the anti-Hortensius manuscript under different Latin titles. He decided that the only way forward was to make the observations with is own eyes:

> Soon after the commencement of my astronomical studies, and whilst

> preparing for practical observation, I computed the Ephemerides of several years, from the continuous tables of Lansberg. Having followed up the task with unceasing perseverance, and having arrived at the point of its completion, the very erroneous calculation of these tables, then detected, convinced me that an astronomer might better engaged upon a better work. Accordingly I broke off the useless computation, and resolved for the future with my own eyes to observe the positions of the stars within the heavens ...[11]

Others had also found Lansberg and Hortensius to be in the wrong. Horrocks was very sound in his criticisms. What he wrote was not constructive or creative astronomy, but it was a necessary criticism to expose the errors in the current theories. His next papers were a little more constructive in that they vindicated the astronomy of Kepler, but they contained no original observation or theory. This is not surprising when we consider that the only observing Horrocks had done at this point in time was to measure the positions of the planets. The problem of the motion of the planets had been solved by Kepler, based on a lifetime of observations by Tycho Brahe, and it therefore seems impossible that Horrocks or Crabtree could add anything new to the findings.

It is hardly surprising that Horrocks's criticisms never got as far as publication during his lifetime. But for a short time he was happy. The cosmology of Copernicus and Kepler was obviously correct. He was convinced that the planets moved in ellipses and not in circles. Both he and Crabtree had repeated and confirmed the telescopic work of Galileo. They had seen the moons of Jupiter and the phases of Venus. Galileo had observed the planet Saturn through the telescope and had discovered some very strange effects. His first drawings showed what appeared to be wings or large moons on either side of the planet. The wing effect did not remain constant. A few months later, the phenomena disappeared altogether when the rings were edge-on to the Earth. It was nearly two generations after Galileo when Cassini viewed Saturn with a telescope with sufficient power to resolve the ring system around it. When Jeremiah Horrocks turned his attention to Jupiter and Saturn, he, too, made a new discovery. It was not as spectacular as new moons or ring systems, but it was very significant.

Having satisfied himself that Mars, Venus and Mercury were in the places predicted by the *Rudolphine Tables*, Horrocks found the same was not true of Jupiter and Saturn. Jupiter was gaining on the predicted position given by Kepler's laws, but Saturn seemed to be losing about one minute of arc every ten years. It was a trivial amount, but Horrocks was able to detect it. The two giant planets seemed to exert some strange attractive influence on each other.

CHAPTER SIX

Gravitation and Mechanics

On 14 May 1638, Jeremiah Horrocks wrote to William Crabtree. He had acquired a new telescope and had used it to study the planet Jupiter. His findings were very similar to those of Galileo:

> I have finally obtained a more accurate telescope. I have been observing on various occasions the satellites of Jupiter which sometimes are more, sometimes fewer in number.
>
> I rarely observe Jupiter without seeing one or more satellites at the same time.
>
> 1638. 7 March, 11.30 pm. I saw one like this. *O
>
> The distance of the satellite from Jupiter was less than Jupiter's diameter, to the east. I have not seen anything yet more than 6 or 7 diameters away.
>
> 12 May, 10 pm. I saw 2 satellites like this. O * *
> 13 May, 9.30 pm. 3 in this configuration * O * *
> 13 May, 11 pm. 4 in this configuration * O * * *
>
> I have other observations but I have selected these as a sample so you can see in what configuration they are normally to be seen, and all of them are almost on the same latitude as Jupiter. I dug out a tube about an inch longer than what I use when I observe objects on the earth. In this way, Jupiter appears as a round body, just like the Moon except for the rays; and the little stars themselves nearby appear brighter.[1]

Jeremiah Horrocks was well pleased with his new telescope. No longer did man need to strain his eyes to view the stars. The telescope seemed to draw them down from heaven and bring them closer to the earth. The 'wondrous tube by art invented' showed many things in the skies which were not visible to the naked eye. He was so pleased that he wrote an ode to his marvellous new astronomical instrument.

> Divine the hand which to Urania's power
> Triumphant raised the trophy, which on man
> Hath first bestowed the wondrous tube by art
> Invented, and in noble daring taught
> His mortal eyes to scan the furthest heavens.
> Whether he seek the solar path to trace,
> Or watch the nightly wanderings of the Moon
> Whilst at her fullest splendour, no such guide
> From Jove was ever sent, no aid like this
> In brightest light such mysteries to display;
> Nor longer now shall man with straining eye
> In vain attempt to seize the stars. Blest with this
> Thou shalt draw down the Moon from heaven, and give
> Our earth to the celestial spheres, and fix
> Each orb in its own ordered place to run
> Its course sublime in strict analogy.

He saw the mountains and the oceans on the face of the Moon and he knew that he was looking at another world. He saw the spots on the surface of the Sun. He saw the sights in the heavens which Galileo had described. The telescope brought the whole universe of the night sky closer to him. And it helped to confirm his belief that the Earth held second place below the Sun:

> For whilst thou see'st the lunar disc display
> Such rocks and ocean-depths unfathomable,
> What powers prevent thy sight of worlds celestial
> From tracing all their semblance to this earth?
> This hand divine, right bold Copernicus,
> Supplies fresh arms to vindicate thy cause,
> Supporting thee who dared to make the worlds
> Revolve by laws unchangeable, it clothes
> The hosts of heaven with earthly forms, and bids
> The earth itself to claim the second place

Below the Sun, a rival to the stars
That hold their stations in the realms of space.

Forbidding more the senseless crowd to rule
O'er minds whose high-aspiring thoughts shall soon
Surpass the utmost bounds of ancient lore,
Its powers disperse the troop that know no rule
But texts too vainly taught by him who gave
Such lasting honors to Stagira's name;
They tear to shreds a thousand fancied laws
That truth deface like spots upon the Sun,
And send the tomes that else might lead astray
A fitting present to the moths and worms.

He turned his telescope on his favourite planet: Venus, the goddess of love. But he knew she did not shine by her own light any more than did the Earth or the Moon. Seen through the telescope, Venus was a miniature version of the Moon, with her phases and crescents: the final proof that the Sun and not the Earth was at the centre of the universe:

This prying tube too shews fair Venus' form
Clad in the vestments of her borrowed light,
While the unworthy fraud her crescent horn
Betrays. Though bosomed in the solar beams
And by their blaze o'erpowered, it brings to view
Hermes and Venus from concealed retreats;
With daring gaze it penetrates the veil
Which shrouds the mighty ruler of the skies,
And searches all his secret laws. O! power
Alone that rivalest Promethean deeds!
Lo, the sure guide to truth's ingenuous sons!
Where'er the zeal of youth shall scan the heavens,
O may they cherish thee above the blind
Conceits of men, and the wild sea of error
Learning the marvels of this mighty Tube![2]

Horrocks began to wonder about the mechanics of the heavens.

What were the forces which drove the planets around the Sun? What were their nature and what were their laws? The Sun seemed to have a strange attraction for the planets, but what was it that kept them perpetually moving around him in their orbits? The only known forces capable of doing this, and which seemed to act at a distance, were magnetism and electricity. Horrocks had read William Gilbert's thesis on magnetism and he knew the principle of the lodestone and the curious effect it had on metallic objects. He knew of the existence of the Earth's magnetic field. Gilbert also wrote about electrostatic force. He knew that when a rod of amber was rubbed, it could attract and pick up small particles. The electrostatic force was seen as a small, localized phenomenon. It did not seem to be appropriate to the heavens. But the magnetic force of the Earth could be demonstrated to exist throughout the planet. Could this be the force which kept the planets in motion?

William Gilbert suggested that the Earth had certain magnetic fibres which extended from the North to the South Pole. He likened this magnetic field to the grain in a piece of wood. Horrocks, too, believed that the magnetic fibres were like the grain in the wood, though he thought they were not straight, but irregular in some places, like knots in wood. The grain followed a general direction but in a sinuous, crooked line:

> Unto this grain the needle of the compass doth bear an exact sympathy, and where it doth bende just north the compass varys nothing at all; but where the fibre do ly anyway but north and south, the compasse following them doth vary from the true Azimuth. This conceit of mine is confirmed by the greatness of the variation about the pole, more than neer the equinoctiall, for there (at the pole) the deviation of these fibres [from a] streight line, though it be no greater than at the equinoctial, yet will make a greater angle with ye meridian, because they are nearer ye center, wch is ye pole.[3]

He thought that the magnetic fibres somehow regulated the rotation of the Earth, and that the irregularities could cause unequal motion as the Earth rotated about its axis. He had to reject this idea, however, when he discovered that Tycho Brahe had used very accurate hour glasses to try to discover any irregularity in the Earth's rotation. No variation had been detected by Tycho. Horrocks decided that the matter needed further investigation, but it was 'a thorny matter and I must confesse I know not well what to make of it'.

Kepler had also studied Gilbert's magnetic fibres and he suggested that they were the reason why the planets seemed to be alternately attracted and repelled by the Sun. He thought the circle was the natural path of a planet, but that the actual path, the ellipse, took the planet inside and outside the natural circle, according to the attraction or repulsion of the magnetic field around the Sun. Horrocks was able to show that Kepler's reasoning was in error. The Earth's magnetic field followed the daily rotation of the Earth. Kepler's theory therefore required a concentric inner shell to hold the magnet in a constant orientation to the Sun:

> Therefore he [Kepler] is forced to say yt there is wth in the globe of the Earth, another globe separated from ye outward shell, & not turned about wth ye dayly revolution of ye Earths outward part, but remaining unmoved: and in ye moon & other planets he imagineth the same, besides a third globe wthin ye second to cause their Latitude. Now methinks these are harsh and unlikely conceits & meer shifts. Himselfe confesseth it a strait that he is forced unto, and leaves it as an uncertain & doubtful thing. It is true indeed that he doth sufficiently stop ye mouth of the antients, but it is no strengthning of his own cause, to shew the weakness of anothers,
>
> *Solamen miserum socios habuisse*
> [a poor consolation to have had friends][4]

Horrocks argued that the Sun somehow managed to carry the whole body of the planet around itself and that this required a far greater force than that required to turn a globe on its axis. The lodestone, when it was acting on a compass needle, did not carry the needle around it in the air, but easily turned the friendly point of the needle towards itself and so attracted the needle as a whole. But Horrocks could not find a way of explaining the motion of the planets in terms of a magnetic field. 'How can ye Sun carry about ye planet when ye reverse part is toward it, for if it carry it about, it should move to it, if it expell it late: how can it both love & hate it at once', he wrote. 'I must have another cause, of that ovall figure which it is most certain all ye planets do affect. This will not satisfy me.'[5] He decided that the mysterious force by which the Sun drove the planets was not magnetism but some other strange entity. A spinning stone fell to the earth. The trajectory of the stone did not seem to vary as the stone rotated. He reasoned that if the force on the stone was magnetic, then the direction of the force on it would vary as the stone turned in the air.

> There are in ye body of ye planet fibrae magneticae which are the cause of its motion, but they are not any such corporeall things as Kepler makes them, but all one wth that vertue which the Sun infuse into them: as when the hand doth cast a stone, it hath a vertue infused of flying the way that it is cast; it is no antipathy between ye hand & ye stone, and therefore though ye stone turn in the flying, & on its side so have not ye same part always towards the hand, yet wee need not to imagine that there is another stone wthin, that doth not turn about wth ye outer part; but the hand doth infuse such a virtue into it of moving in a streight line as that ye turning of ye material stone cannot alter the course of the stones way, nor change those spiritual fibres (for so I will call them) out of their parallell position. And ye same is the reason of any body let fall from on high; the attractive power of ye Earth doth direct the virtuall force therof, into a streight line directed toward itselfe, neyther can ye materiall and corporall substance, anything alter ye internal form, wch ye Earth doth infuse into it.[6]

He was getting closer to the cause. He considered a theory whereby the Sun's rays attracted the planet, but he had problems explaining how these rays could cause the tangential motion of the planet. He proposed a system whereby the tangential motion was caused by the rotation of the Sun, the rays sweeping the planets with them as the Sun rotated. He appealed to his Puritanical upbringing to compare the Sun to God and the planets to people who were drawn or repelled from the true faith.

These ideas on gravitation are philosophical and theoretical. But Jeremiah Horrocks was also a very practical astronomer. He was busy

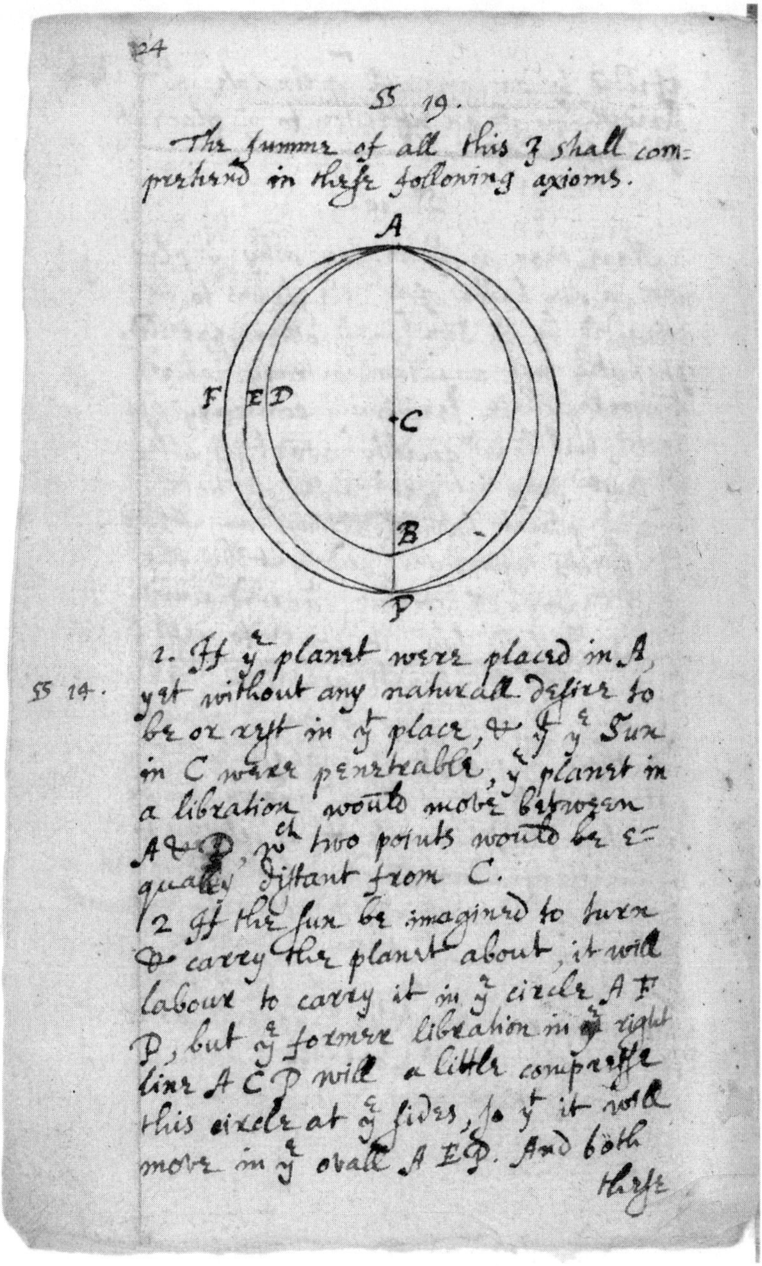

24

§ 19.

The summe of all this I shall com-
prehend in these following axioms.

A

F E D

C

B

P

1. If ye planet were placed in A,
§ 14. yet without any naturall desire to
be or rest in yt place, & if ye Sun
in C were penetrable, ye planet in
a libration would move between
A & P, wch two points would be e-
qually distant from C.
2 If the sun be imagined to turn
& carry the planet about, it will
labour to carry it in ye circle A F
P, but ye former libration in ye right
line A C P will a little compresse
this circle at ye sides, so yt it will
move in ye ovall A E P. And both
these

FIG. 6. Sheet from 'Philosophical Exercises'
Page from Horrocks's notebooks, with a diagram showing the motion of a planet around the Sun.

observing the skies and taking careful measurements of the positions of the planets. At this time – in fact, for many years before he was born – the planet Jupiter was constantly gaining on the position predicted by Kepler's laws. The planet Saturn was constantly losing. The amount was minuscule, but with his meticulous measurements Horrocks was able to detect a real discrepancy. He suggested that the parameters of the orbit of Jupiter might be corrected by adding 1° 30' to the aphelion, and 2' to the mean longitude, but this on its own would not correct the anomaly. The figures would not remain constant. By comparing his own measurements with those of Tycho Brahe, he deduced that the error was increasing by about 1 minute of arc every ten years. In the case of Saturn, he proposed to subtract 4 minutes from the planet's mean longitude as measured by Tycho at the beginning of the century. At first, he was puzzled and annoyed at these errors, small as they were. But when he had convinced himself that the effect was real, he consoled himself with the knowledge that he was in all probability the first person to discover it. He asked William Crabtree to make frequent observations of the two planets, to help verify his findings and so that they could calculate the necessary correction to the figures in the *Rudolphine Tables*. From remarks in Horrocks's letters to Crabtree respecting the periods of the two planets (i.e. the time taken for them to orbit the Sun), he appears to have reached the conclusion that the anomaly occurred once in every orbit. He was very close to the true reason for the errors, which was of course the gravitational attraction between the two giant planets.

At about the same time as he was studying Jupiter and Saturn, Horrocks began to turn his attention to the nature and movement of comets. The comets appeared at random intervals and were sometimes very bright in the night sky. For centuries they were supposed by astrologers to be harbingers of evil. Some astronomers thought that the comets were close to the Earth, inside the orbit of the Moon, like the meteors and shooting stars which burned up in the Earth's atmosphere, but this theory was easy to disprove when it became obvious that astronomers in different countries saw the comets against the same background stars. Horrocks obtained a copy of Tycho's treatise on comets. He discovered that Tycho's measurements showed some comets to be closer to the Sun than the orbits of Venus and Mercury. He was puzzled by the nature of the comets: they were obviously neither planets nor stars. Tycho had done little to calculate their path around the Sun, and Horrocks began to speculate on the shape of their orbits.

At first he formulated the theory that the comets were propelled out of the Sun itself in straight lines. This was in agreement with Kepler's theory, and it was suggested by the marked elongation of their orbits. He wrote to Crabtree, expressing some of his thoughts and asking whether his friend had any observations on comets:

> At the end, he [Longomontanus] has an appendix on comets and he differs as between the two observations of 1607, 1618 AD. The movement of the earth will be demonstrated by both of these, especially the former (unless I am much mistaken), and at the same time it will be shown that comets come forth straight from the Sun itself, moving ever more slowly the further they have travelled from the Sun, until, having become stationary, they at length return again with increased velocity like a curve in a circle, while deviating a little the speed with which the Sun rotates. I would like you to communicate to me all the observations on comets that you have, and certainly of those that are in the progymnasmatis of Tycho, the locations in longitude and latitude, or at least the main points of data, so that after quietly evaluating everything I can set out my assessment to you.[7]

Horrocks was able to show that as the comets receded from the Sun, their velocity diminished, and that they moved faster as they travelled closer to it. Their pattern of motion was therefore similar to that of the planets. He became convinced that the paths of the comets were not straight lines at all, but curves of some kind. He actually stated at one point that they moved in closed curves which return into themselves, and he illustrated his theory with a diagram for the comet of 1577. This brilliant deduction was typical of Horrocks and it was far ahead of its time. He pre-empted Edmund Halley by arriving at the conclusion that comets revolve 'in an elliptical figure or near it'. He recognized that the Sun was the moving force behind the comets, but he failed to make the final step. His diagram showed the comets originating at a point on the Sun rather than orbiting around it.

To later generations, brought up on Newtonian mechanics, it seems natural that an attractive force directed towards a massive body like the Sun can produce motion in a closed orbit around it. Before Newton, however, the notions of force and acceleration were not properly defined. Force and inertia were frequently confused. It was obvious that the Sun provided the moving force for the planets, but the strangest ideas were put forward to try to explain the mysterious effect it had.

Kepler's laws were almost perfect for predicting the planetary positions, and, as far as the astrologers were concerned, they solved the problem of casting horoscopes. But they were no more than empirical models. Kepler's efforts to try to explain them in terms of forces and other physical phenomena were very confused. Kepler thought that the force which moved the planets was applied in the direction of their motion. It seemed obvious to him that a force directed towards the Sun itself would cause the planet eventually to fall into the Sun. Kepler supposed the planets to be whirled round by the action of magnetic fibres, by which, as he thought, a mutual influence was exercised similar to that of the poles of a lodestone or a magnet. Much as he admired Kepler, Horrocks begged to differ from him in religion as well as some of his astronomy:

> Keplers astronomy differs from mine, as his religion: He gives the planets a divers nature (good and bad) that they may eyther come to the Sun or fly away at their pleasure, or at least (as his second thoughts are) so dispose themselves (in spite of all the suns magneticall power) that the sun is bound to attract or expell them, according to that position, which themselves defend against all the suns labouring to incline the fibres. I, on the contrary, make the planet naturally to be averse from the sun, and desirous to rest in its owne place, caused by a materiall dulnes naturally opposite to motion, and averse from the sun, without eyther power or will to move to the sun of itselfe. But then the sun by its rays attracts, and by its circumferentiall revolution carrys about the unwilling planet, conquering that naturall selfe rest that is in it, yet not so far but that the

> planet doth much abate and weaken this force of the sun, as is largely disputed afore.[8]

Horrocks postulated that the laws of nature were everywhere the same. This alone was a great step forward in philosophy, for most people believed that the heavens, which enjoyed perpetual motion, were governed by different laws from those of the Earth, where everything ran down to a halt because of friction and air resistance. He gave as an example the trajectory of a stone thrown from a point on the Earth. We know that without the air resistance the trajectory of the stone would be a parabola, but Horrocks stated that it was an ellipse exactly like the orbits of the planets. A little thought will show that Horrocks was correct: the parabola is the limiting case of a very long ellipse with the centre of the Earth at one focus. Horrocks was thinking of the Earth as a sphere, not a flat surface. He claimed that when the stone was in the air, the rotation of the Earth did not impede its progress. In this analogy, to which he refers more than once, we have his explanation of celestial motion as a combination of tangential motion and attractive forces.

> If a stone be thrown obliquely into the air, its movement is governed by the impulse imparted to it by the hand, together with the attractive power of the earth. In obedience to these two influences, instead of tending in its fall directly towards the centre, it preserves whilst descending the same angle at which it arose; and if its progress were not interrupted by the earth's surface there is little doubt that it would revolve unceasingly in an elliptical orbit with the centre in the lower focus.[9]

This conclusion was the equivalent of Newton's thoughts about the apple. Where Newton speculated as to whether or not the force on the apple was the same as the force on the Moon, Horrocks claimed that the motion of the stone attracted by the Earth was in principle the same as the motion of the Moon around the Earth.

Horrocks arrived at a general law. Whatton, writing in the nineteenth century with the advantage of knowing Newton's laws, explained it as follows:

> When two spheres are mutually attracted, if not prevented by foreign influences, their straight paths are deflected into curves concave to each other, and corresponding with one of the sections of a cone, according to the velocity of the revolving body. Thus if a sphere were projected by an independent power, as the planets were when launched forth from the Creator's hand, it would move forward in a right line for ever, unless attracted from it by an extraneous force; for instance, the earth would preserve a perfectly straight course whilst permitted to do so, but coming within the sun's influence, it is induced to deviate from the direction originally impressed upon it. Now if the velocity with which the revolving body is impelled be equal to what it would acquire by falling through half the radius of a circle described from the centre of deflection, its orbit will be circular; but if it be less than that quantity, its path becomes elliptical. This law was subsequently expanded by Sir Isaac Newton into the great principle of gravitation. As is well known, he concluded that the power which causes a body to fall to the earth, is of the same nature as that which retains the planets in their orbits; and he pursued this discovery, until he finally evolved an expression to which the phenomena of all the celestial movements may be confidently referred. Whilst thus engaged, he derived important assistance from the writings of Horrocks, who, by his sagacious application of projectile to celestial motion, has gained a distinguished place amongst those whose labours have contributed to the establishment of the true system of the universe.[10]

To put Horrocks's ideas into perspective, he did not postulate the universal law of gravitation and he did not suggest an inverse square law. Even if he had arrived at the idea of an inverse square law, he could not have demonstrated that the motion of the planets could be deduced from such a law. The rigorous mathematics required to prove this – namely the fluxions, or what we now call the calculus – had simply not been invented in his time. Others after Horrocks but before Newton had speculated on the inverse square law of gravitation, but it was Newton's mathematical genius which enabled him to demonstrate that the motions of the planets could be deduced from the one fundamental law of gravity. None the less, Horrocks showed an amazing grasp of the problem. In a passage which could have been written by Newton himself, Horrocks described in his own words the principle of a planet's orbit:

> It is surely conceded by all that the motion of the planetary bodies is neither perfectly circular, nor perfectly uniform; for observations shew, beyond dispute, that the figure of the planetary orbits is elliptical or oval, and different from a circle: and the motion of a body in this ellipse is irregular being increased or diminished according to its distance from the Sun. Physical causes are not wanting to shew that this movement is described by a sort of geometrical necessity. We may satisfy ourselves of the truth of this by an appeal to nature; for as a planet is moved by a magnetic impulse, why may not the same principle be exercised in other ways? A weight is thrown into the air: at first it rises quickly, then moves slowly, until at length it is stationary, and falls back to the earth with a velocity which continually increases. It thus describes a libratory movement. This movement arises from the impetus in a right line which has been imparted to it by your hand, together with the magnetic influence of the earth, which attracts all heavy things to itself, as a loadstone does iron. There is no need to dream of circles in the air, and I know not what, when we have the natural cause before our eyes; and as regards the motion of the planets which are subject to similar influences, what reason, I ask, is there to barter an explanation, the truth of which is confirmed by so many examples in nature, for a fictitious dream of circles?[11]

Isaac Newton, when he came to describe the same problem of comparing the motion of the projectile with the motion of the Moon, wrote a passage which is very similar. It shows very clearly the parallel thought processes by which the two astronomers arrived at their similar conclusions:

> Just as by the force of gravity a projectile might describe an orbit, and revolve round the whole earth; so the moon, either by the force of gravity if it is endued with gravity, or by any other force urging it towards the earth, may be continually drawn thereto from a rectilineal path, and turned into her present orbit; and without such a force she cannot be retained in her orbit. If the force were less than it is, it would not cause her to deviate from it rectilineal course sufficiently: if it were greater it

> would cause her to deviate too much, and draw her from her orbit towards the earth. It is therefore required to be of an exact amount; and it is the business of mathematicians to find the force which can accurately retain a body with a given velocity in any given orbit; and in like manner to find the curvilineal path into which a body going forth with a given velocity from any given place is turned from its rectilineal way by a given force.[12]

Thus Horrocks was very close to the truth with his idea of a projectile. Where his theory failed was that he could not grasp the idea that a radial force alone could carry the planets around the Sun. He thought that a tangential force was also required and that this was somehow provided by the rotation of the Sun. He tried to find physical models which could be used to illustrate his ideas and he suggested a pendulum analogy to simulate the motion of the planets. The pendulum, consisting of a spherical lead bob on a string, was suspended from a fixed point. If the pendulum was pulled from its rest position and released, it would move in a plane like the pendulum of a clock. If it were given a tangential impulse, however, it would trace out an oval figure similar to the orbit of a planet about the Sun. If the tangential impulse was exactly right, then the motion would be in a circle, but in general the curve traced out would be an ellipse. Horrocks knew that the analogy was not very good. The force was directed to the centre of the ellipse rather than the focus, and Kepler's second law regarding the radius vector did not hold. He introduced the idea of a wind to force the ellipse off-centre. This could be easily demonstrated by means of a pair of bellows. The introduction of a shear force brought the model closer to the truth but it was still a long way away from the rigorous mathematical approach applied by Newton fifty years later. Horrocks used the pendulum analogy again when he came to model the motion of the Moon. On this second occasion, the wind was used to simulate the gravity of the Sun acting on the Moon.

The pendulum experiment was repeated many years later, in the 1660s, by Robert Hooke at the Royal Society. Hooke hung a brass ball, representing the Moon, by a long cord from the ceiling. He gave it a tangential impulse to set it in motion to simulate the elliptical orbit. In the centre of this 'orbit' he placed a larger ball, representing the Earth. By using a pair of bellows, he simulated the gravity of the Sun. The Fellows of the Royal Society were fascinated to see that the long axis of the ellipse – in fact, the whole of the ellipse – rotated longitudinally about the stationary ball representing the Earth.

The motion of the Moon was the next theoretical problem to which Horrocks applied his mind. He was to spend more than two years studying the problem. His ideas developed gradually over this period and he was still working on it at the time of his death. He recognized that the orbit of the Moon was very complex compared to the orbits of the planets. The orbit itself was not fixed. It rotated or precessed about the Earth and one of the cycles in the precession had a period of one year. This implied that although the Earth was the major factor governing the motion of the Moon, the motion was also influenced by the Sun.

In a letter of 23 November 1638 to William Crabtree, Horrocks developed his ideas on the Moon. He stated that 'motion in an oval figure may be deduced from natural principles'. In the same letter he states that the motion of the planets is caused firstly by the attraction of the Sun but secondly from a force due to the rotation of the Sun which keeps the planets in orbit and prevents them from falling into the Sun. When he came to study the Moon, the nineteen-year-old Horrocks made a case for an attractive force and a turning force of the Sun superimposed on an attractive force and a turning force from the Earth.[13]

Horrocks had made two important contributions to the theory of gravitation. One was the discovery of the gravitational attraction between the two giant planets Jupiter and Saturn. The other was the fact that the Moon's motion was influenced by the gravity of the Sun as well as that of the Earth. Horrocks was not the first to realize that the Sun influenced the Moon's motion, but his subsequent work showed that he had a strong claim to understand this phenomenon better than any astronomer before him.

It would be possible to develop a system of celestial mechanics based on the ideas put forward by Jeremiah Horrocks. Any system, including that developed by Newton, has to resolve the motion into radial and tangential components. The great beauty and simplicity of Newton's system is that everything is developed from three laws of motion and the single principle of the Law of Gravitation. If Horrocks had managed to develop his system further, it would not have been as elegant as that of Newton. He was faced with two great problems. One was that the familiar, well-defined terms of mechanics – such as force, momentum, energy and inertia – were not defined in Horrocks's time, so all his writings are subject to insuperable problems of semantics. The other problem, already mentioned, was the mathematics of the calculus which was needed to put his ideas into a rigorous form. Horrocks's college friend John Wallis did make a contribution to the origins of the calculus, but even this came after the death of Horrocks. He seems to have been aware of the work on force and acceleration done by Galileo, but he was still an impossibly long way away from anything as rigorous as a mathematically based theory of celestial mechanics. What Horrocks did achieve, however, was the next best thing. When he applied his mind to the problem of the Moon's motion, he knew that if he could produce a model which could predict the position of the Moon with a reasonable degree of accuracy, then this would throw some light on the basic laws behind the mechanics. His letters to Crabtree in 1638 show that he had already put his mind to the problem at the age of nineteen. Two years later, by the time he was twenty-one, his theory was the most successful ever developed to that point in time and it remained so for nearly a hundred years.

CHAPTER SEVEN

Transit of Venus

After leaving university, Jeremiah Horrocks had returned to Toxteth, where he worked with his father in the family business. There is every reason to believe that his family gave him the support and patronage he needed to follow his love of astronomy, but as a young man living with his parents he soon got the urge to leave the family home and to move on. After about three years in Toxteth, it was natural that Jeremiah should want to work for somebody other than his father.

In the summer of 1639, having almost reached the age of twenty-one, Jeremiah left home. He did not travel far, only about eighteen miles along the coast to the parish of Hoole. There is a long and enduring tradition that Horrocks went to Hoole as curate of the chapel of St Michael and All Angels. The Revd A. B. Whatton, Horrocks's Victorian biographer, was convinced that Horrocks was actually ordained and gives him the title of the 'Rev. Jeremiah Horrox'. He quotes as the source of his information a treatise by the astrologer John Gadbury which he claims puts the matter 'fortunately beyond conjecture'.

> Ephemerides of the planetary motions, eclipses, conjunctions, and aspects for fifty years to come, calculated from the British tables, composed first by the Reverend Mr. Horrox, and first published by Jeremy Shakerley.[1]

Later researchers have been unable to find the document quoted by Whatton, but even if it is rediscovered, it does not give direct proof of Horrocks's ordination. The records of the Church of England show nothing to support the hypothesis that he was ordained. Both

Gaythorpe[2] and Chapman[3] argue that Jeremiah Horrocks was too young to be ordained: he was under twenty-one when he arrived at Hoole and the minimum age was twenty-four. The little chapel at Hoole was built in 1628 and at that time it was not a parish church but a chapel of ease to the mother church of Croston within which parish it lay. The church records show that a curate called Robert Fogg was appointed in 1632. His name also appears in 1639, so it seems that he was the curate of Hoole for most of the 1630s. Robert Fogg went on to become the first rector when the chapel was awarded the status of a parish church in 1641. This disposes of the theory that Jeremiah Horrocks was the curate.

Horrocks, with his university education and his religious views, may well have contemplated a career in the Church, provided of course that it still allowed him time for his astronomy and provided also that he would be allowed to practise his Puritan beliefs. The church of St Michael and All Angels at Hoole has every right to its claim for Jeremiah Horrocks. He was a devoted Christian and he was almost certainly an active member of the congregation. He lived at a time when the majority of people had strong religious beliefs, and although his viewpoint was shaped by almost exclusive contact with the Puritan movement, this did not exclude him from involvement with the Church of England. Something about his religious leanings emerges from his writings, where he occasionally digresses into religion. In one passage at the conclusion of his 'Philosophical Exercises', he compares the attractive influence of the Sun to that of God. He leaves us a little confused about his beliefs and all it does is show that he was unorthodox:

> I will confesse myselfe not equally composed of good and bad, that myselfe may give eyther flesh or spirit the upper hand; but rather wholly desirous to rest in my selfe, wholly averse from God, and therefore justly deserve (as the fixed stars from the Sun) to be blown away from God in infinitum, but that God by his Sons taking on him mans nature, and the undeserved inspirition of his spirit, doth quicken this dulnes, nay deadnes of my nature, yet still; ah me! how doth it choke and weaken those operations!

> If any one thinke all this but an idle conceit, I must tell him he doth too rashly deride that booke of creation, that voyce of the heavens which is heard in all the world, and wherein without question God hath instamped more mysterys than the lazy witts of men, more ready to slight than amend any speculation, are ordinarily aware of. Shall we thinke that he who was content to shadow out these mysterys with the poor blood of buls and goats, will disdain to have them typified in the more glorious bodys of the stars and motions of the heavens; which David accounted such cleare Emblems of Gods glory that he goes from speaking of the light of the Sun, unto Gods law, as if the subject were still the same, without any conclusion to the first, or introduction to the latter.
>
> For my part I must ever think that God created all other things, as well as man in his own image, and that the nature of all things is one, as God is one, and therefore an harmonicall agreeing of the causes of all things, if demonstrated, were the quintessence of most truly naturall philosophy.[4]

At another point in his 'Philosophical Exercises', Horrocks makes it very clear that he believes the universe to be governed by the unchangeable laws of God. He postulated that the laws of God could be discovered by man but that there would always remain some anomalies, put there by God, which would put the final truth beyond the reach of man:

> Besides it may seem unfit that any creature should so far participate of infinitenese and perfection, as by any such divine perpetuity to forget that it were a created excellency: It is ye property of God to be ye same yesterday, to day & so for ever. Lastly let him that will needs stoically defend this immutable regularity, show it, by such Tables as shall truly be called perpetuall: let him silence all further disputes by claiming these refactorys in unchangable laws; and not fondly credit his Aristotles word so far, as rashly to censure their skill, who have in vain attempted it, himselfe in ye meanwhile content to be an idle, and therfore censorious spectator, rather then by tryall be forced to a confession of ye knottines of such a taske,
>
> *Tum sciet ignipedum vires experitus equotum*
> *Non meruisse necem qui non bene rexerit illos.*

> [He who has himself tried the strength of those fiery-footed steeds will know that the one who failed to guide them well did not deserve death.][5]

The Latin quotation is from Ovid's *Metamorphoses*, II, lines 392–3. The lines are spoken by Phoebus, the driver of the Sun's chariot, who feels that others do not appreciate the difficulties of his job.

The seventeenth-century scientists were nearly all devout Christians, but there were so few scientists in Horrocks's time that we need to look later in the century to find the evidence. John Flamsteed was a very devout believer and he held the living of Burstow in Surrey. Robert Hooke's father was the incumbent of Freshwater in the Isle of Wight and Christopher Wren's father held the living of East Knoyle in Wiltshire. The university system was almost exclusively geared towards theology. Nobody in England could enrol for higher education unless they declared themselves to be a devout believer. John Wilkins of the Royal Society became the bishop of Chester and Seth Ward became the bishop of Salisbury. Isaac Newton was the son of a farmer, but his stepfather and his uncle were clerics and he wrote millions of words on the interpretation of the scriptures, though he declined to take holy orders himself. The exception to the rule was Edmund Halley, who in later life upset Flamsteed with his atheistic beliefs. Part of the problem with Halley was that he had commanded a scientific voyage to the South Atlantic and when he returned he could swear an oath as well as any seaman.

Horrocks described his move to Hoole as being only eighteen miles north of Liverpool. This is an astronomer's distance, the difference in latitudes being about 18 minutes of arc. For those who were unable to fly, the journey was a little further. A muddy country road ran northwards from Liverpool through Aughton, Ormskirk and Rufford. It was hardly a well-maintained highway. Later in the century, when Celia Fiennes travelled that way, she complained that she could have travelled twice the distance in the same time on the roads around London. There was another road, however, which linked the coastal communities of South Lancashire: a fine, broad highway of firm sand. The beach was so firm that at Liverpool on high days and holidays it was used for horse-racing. The coastal communities used the beach as their highway. It ran northwards from Liverpool to Crosby, around Formby Point, and then became Ainsdale and Birkdale Beaches, leading to Churchtown in the parish of North Meols. It was an invigorating ride for a young man of Jeremiah's age for a distance of about eighteen miles. North of Churchtown, the sand-hills petered out to be replaced by the mudflats of the Ribble Estuary. Inland lay a great shallow mere called Marton Mere, about six miles in diameter. It was the greatest expanse of water in the county. The mere drained into the river Douglas near Rufford. The road from Churchtown crossed over the river Douglas and entered the scattered rural community of Hoole.

The more remote coastal communities of Lancashire were seldom visited by outsiders, but fortunately there is one excellent description from the pen of Dr James of Oxford. He described a Lancashire coastal village in 1635, very close to the time when Horrocks passed through on his way from Toxteth to Hoole, and he captures the atmosphere of isolation very accurately. Dr James's description is an account of Churchtown, now a suburb of Southport but in the seventeenth century the main centre of population in the rural parish of North Meols. Dr James describes the fishing community and puts his impressions into rhyme, although this is not obvious on a first reading. He speaks with an old fisherman, probably a veteran of the Spanish Armada, who had served in the Elizabethan navy under the Earl of Essex.

Ye guize
Of those chaffe sands which doe in mountains rize
On shore is pleasure to behold, which Hoes [Hawes or sandhills]
Are called in Worold [the Wirral]; windie tempest blowes
Them up in heapes; 'tis past intelligence
With me how seas do reverence
Upon ye sands; but sands and beach and peobles are
Cast up by rowling of ye waves a ware
To make against their deluge. Since the larke

And sheepe within feede lower than ye marke
Of each high flood. Heere through ye washie Sholes
We spye an owlde man wading for ye soles
And flukes and rayes, which the last morning tide
Had stayed in nets, or did at anchor ride
Upon his hooks; him we fetch up and then
To our goodmorrow, 'welcomme gentlemen'
He sayed, and more, 'You gentlemen at ease
Whoe money have and goe where ere you please
Are never quiett; wearye of ye day
You now come hither to drive time away;
Must time be driven? longest day with us
Shutts in too soon, as never tedious
Unto our buisnesse; making mending nett
Preparing hooks and baits wherewith to gett
Cod, whiting, place, upon the sandie shelves
Wherewith to feed the market and our selves.
Happie ould blade, who in his youth had binne
Roving at sea, Where Essex Cales [Cadiz] did winne
So now he lives. If any Bushell will
Live west the world, withoute projecting skill
Of Ermitage, he shall not need to seek
In rocks, or Calve of Man, an ember weeke,
Heere at ye desert Meales he may, unknowne
Bread by his won paines getting, live alone
Without a callot or a page to dress
Or bring bought meat unto his holiness[6]

If Jeremiah Horrocks was not the curate of Hoole, then what motivated him to move from Toxteth to such an isolated spot? It is possible that there was a romantic connection. His youth and his flowery descriptions of his beloved Venus imply that he appreciated the attractions of the opposite sex, but there is no evidence of any particular romantic attachments in his writings or his letters to William Crabtree. He was firmly wedded to his astronomy.

In a survey taken in 1641, the population of Much Hoole was

recorded as 235 adults. The choice of residences was not very great. The vernacular-style farmhouses of Hoole were low and thatched, with smoking peat fires. They usually had pigs and chickens running in and out of the doorway. They were a far cry from the ideal residence for an astronomer. Apart from the church itself, the only building in the village suitable for an observatory was Carr House at Bretherton, the home of the Stones family. It was situated about half a mile away from the church. Hoole is the last place where we might expect to meet merchants from great trading centres like London and Amsterdam, but there is sometimes a tendency to overplay the isolation of the English village. At Carr House, above the doorway is an inscription in tablets of stone which proves the trading connections to be a fact:

> Thomas Stones of London haberdasher and Andrewe Stones of Amsterdam marchant hath builded this howse of their own charges and giveth the same unto their brother John Stones: Ano Domini 1613 Lavs.[7]

The Stones family were evidently haberdashers. There is a large overlap between haberdashery, drapery and cloth dealing. It raises the question of whether Messrs Worthington and Crabtree knew the Stones and whether they had something to do with Jeremiah's move to Hoole. The most plausible theory which has been put forward for the move to Hoole is that Jeremiah Horrocks was employed in some capacity by the Stones family, possibly as a private tutor to their children.

Whatever his employment, Horrocks was still able to indulge in his astronomical studies. When at Hoole, he continued to monitor the motion of the planet Venus. Once again, his meticulous observations did not agree with those of Johannes Kepler, but this time, unlike Saturn and Jupiter, Horrocks did not find fault with the theory, simply the accuracy of Kepler's observation. Kepler had predicted that in 1631 the planet Mercury would pass over the face of the Sun and this event was actually observed by the astronomer Gassendi. Kepler himself tried to make the observation, but much as Horrocks and others admired Kepler's mathematics, he was a complete disaster as an observational astronomer. Kepler claimed to have observed Mercury on the face of the Sun, but he failed to recognize that in reality all he had seen was a sunspot. Kepler also correctly predicted that Venus would transit across the face of the Sun in the same year. The transit of Venus was a much rarer event than that of Mercury, but unfortunately the observation could not be made in Europe because the event took place at night when the Sun was illuminating the far side of the Earth. According to Kepler's calculations, it would be 130 years before this astronomical event could be seen again.

At this time, the main reason for observing an event such as the transit of Venus was astrological rather than astronomical. Conjunctions of the Sun and the planets were seen as signs, rather like a new comet or the star of Bethlehem, which were supposed to indicate that a great event was about to happen on Earth. A good observation allowed the astronomers to correct the orbit of a planet, and this was useful to the astrologers when it came to casting 'more accurate' horoscopes.

Horrocks was delighted to discover that Kepler's estimate of more than a century until the next transit of Venus was very wrong. He began to suspect from his measurements that the transits of Venus occurred in pairs and that another transit was about to take place that very year (1639). He measured the position and motion of Venus whenever he could see the planet and became more and more convinced that it was heading for a conjunction with the Sun. His vigil is recorded in his account of the transit, the Danzig publication of *Venus in Sole Visa*:

> When first I began to attend to this Conjunction, I not only determined myself to watch diligently an appearance so important, but invited others also whom I knew to be interested in astronomy to do the same, in order that the testimony of many observers, should it so happen, might more firmly establish the truth; and especially because if observations were made in different places, our expectations would be less likely to be frustrated by a cloudy sky or any other obstacle. I wrote therefore immediately to my most esteemed friend William Crabtree, a person who has few superiors in mathematical learning, inviting him to be present at this

> Uranian banquet, if the weather permitted, and my letter, which arrived in good time, found him ready to oblige me[8]

In the letter to Crabtree, which survives in the *Opera Posthuma*, he wrote with great excitement to tell of his latest discovery:

> My reason for now writing is to advise you of a remarkable conjunction of the Sun and Venus on the 24th of November, when there will be a transit. As such a thing has not happened for many years past, and will not occur again in this century, I earnestly entreat you to watch attentively with your telescope, in order to observe it as well as you can. Notice particularly the diameter of Venus, which is stated by Kepler to be 7 minutes [of arc], and by Lansberg to be 11 minutes but which I believe to be scarcely greater than 1 minute. If this letter should arrive sufficiently early, I beg you will apprise Mr. Foster of the conjunction, as, in doing so, I am sure you would afford him the greatest pleasure. It is possible that in some places the sky may be cloudy, hence it is much to be desired that this remarkable phenomenon should be observed from different localities since a slight change of numbers in Kepler's tables (which indeed might not be admitted notwithstanding other observations which I know), would alter the time of conjunction and the quantity of the [planet's] latitude: it will be best to look out for it throughout the day, even from the evening before and up to the following morning, if it is not seen in the meanwhile. But I quite think that it will take place on the 24th day [of November][9]

The letter was written on 26 October. It is of interest that Horrocks already suspected that the size of Venus on the Sun's disc would be only about 1 minute of arc, very much smaller than previous estimates. He asked Crabtree to write to Samuel Foster at Gresham College, although it would surely have saved valuable time to write himself. Perhaps the postal service from Hoole was very spasmodic. Crabtree probably obliged, but the information did not reach London in time for Foster to partake in the observation. Jeremiah also wrote to his brother Jonas in Toxteth, asking him to look out for the transit and to make the observation. He wanted to communicate his discovery to the whole of the astronomical world, but there was just not enough time to achieve this before the event.

The year before the transit of Venus, Jeremiah Horrocks had made plans to observe a transit of Mercury from Toxteth. He made the necessary preparations using a plumb line to check his apparatus and he devised a movable pointer to help locate Mercury on the face of the Sun. All his plans for Mercury came to nothing. His calculations were in error and the transit was not visible in England. The preparations for Mercury were not in vain, however, since it meant that he was better prepared when the time came to observe Venus. In a burst of youthful enthusiasm, he rejoiced in his good fortune at having discovered that the event would take place. He eulogized to Gassendi, the only astronomer ever to have observed Mercury on the Sun, about Venus, the Queen of Love:

> Hail! Then, ye eyes that penetrate the inmost recesses of the heavens, and gazing upon the bosom of the Sun with your sight-assissting tube, have dared to point out the spots on that eternal luminary! And thou too, illustrious Gassendi, above all others, hail! Thou who, first and only, didst depict Hermes' changeful orb in hidden congress with the Sun. Well hast thou restored the fallen credit of our ancestors, and triumphed o'er the inconstant Wanderer. Behold thyself, thrice celebrated man! Associated with me, if I may venture to speak, in a like good fortune. Contemplate, I repeat, this most extraordinary phenomenon, never in our time to be seen again! The planet Venus drawn from her seclusion, modestly delineating on the Sun, without disguise, her real magnitude, whilst her disc, at other times so lovely, is here obscured in melancholy gloom; in short, constrained to reveal to us those important truths, which Mercury, on a former occasion, confided to thee.
>
> How admirably are their destinies appointed! How wisely have the decrees of Providence ordered the several purposes of their creation! Thou, a profound Divine, hast honoured the patron of wisdom and learning; whilst I, whose youthful days are scarce complete, have chosen for my theme the Queen of Love, veiled by the shade of Phoebus' light![10]

As the day of the transit approached and Venus moved nearer the glare of the Sun, it became more and more difficult to measure her position in the sky. Horrocks made some calculations based on the data he had available. He discovered that the transit of Venus would be visible through nearly the whole of Italy, France and Spain as well as in Britain, but none of these countries would be able to see the entire continuance of the event, which lasted for several hours. The best place from which to see and observe the whole transit was America. Jeremiah knew as much about the American colonies as most people in England. He knew that it was far too late to send a letter across the Atlantic to his cousin Thomas. He also knew that there was precious little time for astronomical pursuits in Plymouth or Boston, Massachusetts. The Pilgrim Fathers and the early colonists had far more imminent problems than astronomy on their minds. They were concerned about the state of the harvest, the trust of the Indians and the building up of their settlement. It was an exaggeration to say that there was not a single telescope in the whole of America, but Horrocks lamented the loss of the observation with a quotation from Virgil:

> *O fortunatos nimium bona si sua norint!*
> [O how very happy are those if they know their possessions are good]

'Venus!' he wrote,

> what riches dost thou squander on unworthy regions which attempt to repay such favors with gold, the paltry product of their mines. Let these barbarians keep their precious metals to themselves, the incentives to evil which we are content to do without. These rude people would indeed ask from us too much should they deprive us of those celestial riches, the use of which they are not able to comprehend. But let us cease this complaint, O Venus! and attend to thee ere thou dost depart.

His lament was against the American Indians rather than the colonists. In typical Horrocks fashion, he burst into poetry to express his feelings, complaining to Venus that she should display herself to America more than to Europe.

> Why beauteous Queen desert thy votaries here?
> Ah! why from Europe hide that face divine,
> Most meet to be admired? on distant climes

Why scatter riches? or such splendid sights
Why waste on those who cannot prize their value?
Where dost thou madly hasten? Oh! return:
Such barbarous lands can never duly hail
The purer brightness of thy virgin light.
Or rather here remain: secure from harm,
Thy bed we'll strew with all the fairest flowers :
Refresh thy frame, by labors seldom tried,
Too much oppressed; and let that gentle form
Recline in safety on the friendly couch.
But ah! thou fliest! And torn from civil life,
The savage grasp of wild untutored man
Holds thee imprisoned in its rude embrace.
Thou fliest, and we shall never see thee more,
While heaven unpitying scarcely would permit
The rich enjoyment of thy parting smile.
Oh! then farewell thou beauteous Queen! thy sway
May soften nations yet untamed, whose breasts
Bereft of native fury then shall learn
The milder virtues. We with anxious mind
Follow thy latest footsteps here, and far
As thought can carry us; my labors now
Bedeck the monument for future times
Which thou at parting left us. Thy return
Posterity shall witness; years must roll
Away, but then at length the splendid sight
Again shall greet our distant children's eyes.[11]

Horrocks calculated that the transit of Venus would take place on 24 November 1639 and he thought it would begin late in the afternoon. This was the date according to the old-style Julian calendar which was used in England at that time. The date of 4 December is sometimes quoted for the transit because this was the date on the new-style Gregorian calendar, used in Europe but not adopted in England until 1752. It was only four days before the earliest sunset of the year.

Horrocks knew how to arrange his telescope to project an image of the Sun on to a screen. This was the occasion when he claimed that his telescope was 'much more accurate than those generally used'. We know from his description that he owned a Galilean telescope. This had a concave rather than a convex eyepiece, and retained an upright image for normal observation but an inverted image when used for projection. His apartment was too narrow to allow a large projected image, but it was long enough to get a very sharp image of the Sun, about six inches in diameter. He set up his apparatus in readiness to observe the transit. He knew the story of Kepler's abortive attempt to observe Mercury and, so that he would not be guilty of the same error, he made himself familiar with the markings on the face of Sun at that time. He drew a circle of six inches diameter on a piece of card. He then divided the circumference of the circle into 360°. He drew a diameter at the orientation of the ecliptic and he divided it into 120 equal parts. This diagram was, in his opinion, large enough for all practical purposes. He did not think it necessary to carry the subdivisions any further. He knew he could depend upon his eye to judge the fractions of the marked divisions. He adjusted his apparatus so that the image of the Sun should be transmitted perpendicular to his diagram and it exactly filled his described circle. The card was fixed to the telescope by a length of wood so that the distance of the image from the telescope lenses would be fixed to give a sharp focus. This was a necessary precaution so that the whole apparatus could be easily moved to follow the position of the Sun through the sky.

On the first floor of Carr House, above the entrance porch, is a mullioned box window. The window faces south, the five main lights giving a good view in that direction. To left and right are other lights, two on each side, giving restricted views to the east and west. The view to the north is cut off by the wall of the house. There is a similar window on the floor above, but it does not have the side lights of the first-floor window. Horrocks is thought to have observed the transit of Venus from one of these windows. The observations made by Horrocks are just possible from the first-floor window within the restrictions imposed by the stone mullions.[12] Whilst this does not constitute proof that the transit of Venus was observed from Carr

House, it does prove that it was a possible observation point and it is therefore the most likely place. One flaw in the logic is that the first-floor room is one of the best in the house and it was therefore unlikely to be occupied by an employee of the family or even a guest. The room on the second floor would be a more likely room for Jeremiah Horrocks to have occupied. It would have been possible to make all the observations from the second-floor window, except for the final few minutes of the sunset. This does not rule out the second floor as a possible observation point. There is a side window on the landing from which the sunset could have been observed by Horrocks.

From his own calculations, Horrocks did not expect that the transit would take place before three o'clock on the afternoon of 24 November. But it appeared from the tables of others that it might occur somewhat sooner, and in order to avoid the possibility of disappointment, he began to observe about midday on the 23rd. As expected, he saw no sign of the image of Venus on the Sun. The next day, he continued his vigil until he tells us he was 'called away by business of the highest importance, which could not with propriety be neglected'. This phrase has given rise to the legend that he had to go and deliver a sermon just as the transit was about to begin. He tells us nothing else about the business of the highest importance. It was a Sunday and it is a reasonable deduction that he had to perform a Sabbath-day duty of some kind, but the task cannot have been too onerous for he was back at his telescope in just over an hour. There was just enough time to get to the church, help Robert Fogg serve communion and return to his telescope, but his duties may have been no more than a short prayer for the family meal.

When he returned to his room and his telescope, he was overjoyed to see a large, dark, round spot already fully entered upon the image of the Sun. It was without doubt the transit he had been waiting for, the marvellous reward for all his hours of observation. He did not want to be accused of seeing no more than a sunspot, even though it would have put him in the company of Kepler. He would go to great lengths in his treatise to explain that Venus appeared on the Sun's disc as a perfectly circular dark spot. He was overjoyed with the spectacle.

It was an event which he knew had never been seen before in the history of astronomy. He measured the size of the dark spot as accurately as he could and he drew its exact position on his image of the Sun. He drew two more images and recorded the times as 3.15, 3.35 and 3.45. The image moved by one diameter in the first twenty-minute interval, slightly less in the second interval. Then the Sun set over the Ribble Marshes. He knew the value of accurate measurements and he wanted his observation to be as precise as possible. He was working to angles within seconds of arc. He estimated the diameter of Venus as 1' 12" and he estimated his error as 4 or 5 seconds of arc.

At Broughton, Horrocks's friend William Crabtree was also trying to make the observation. Crabtree was very unfortunate with the weather: the skies were overcast for the greater part of the day. He was not able to see the Sun all day and he despaired of ever making an observation. He had decided to give up until a little before sunset, about thirty-five minutes past three, at the same time as Horrocks was making his observation, when the Sun suddenly burst out from behind the clouds. Crabtree rushed into his house and began to observe at once. To his great joy, he saw the rare spectacle of Venus upon the Sun's disc. In a passage which does much to illuminate the personality of both men, Horrocks recorded the feelings of his friend:

> Rapt in contemplation, he stood for some time motionless, scarcely trusting his own senses, through excess of joy; for we astronomers have as it were a womanish disposition, and are overjoyed with trifles and such small matters as scarcely make an impression upon others; a susceptibility which those who will may deride with impunity, even in my own presence, and, if it gratify them, I too will join in the merriment. One thing I request: let no severe Cato be seriously offended with our follies; for, to speak poetically, what young man on Earth would not, like ourselves, fondly admire Venus in conjunction with the Sun. Pulchritudinem divitiis conjunctam? [beauty conjoined with wealth] What youth would not dwell with rapture upon the fair and beautiful face of a lady whose charms derive an additional grace from her fortune?[13]

William Crabtree only saw the spectacle for a very short time, but in the little time he did have, he managed to observe something very similar to that described by Horrocks. Like Horrocks, he made a drawing showing the position of Venus on the face of the Sun. He found that the diameter of Venus was seven parts compared to two hundred parts for the diameter of the Sun. This gave a figure of 1' 3" for the diameter of Venus. It was, as luck would have it, more accurate than the figure arrived at by Jeremiah Horrocks.

Horrocks had also written to his younger brother at Toxteth, 'hoping that he would exert himself on the occasion'. His wording suggests that Jonas was not as enthusiastic as his brother was about astronomy. Although Jonas claimed to have watched carefully all through the day on the 24th, the skies around Liverpool were overcast and he was unable to see anything. Jonas examined the Sun again the following day, when the weather was clearer, but with no better success for Jeremiah had predicted the day of the transit correctly and it was too late. Jeremiah's next words also show something of his open and friendly character. He was ragged for his obsession with astronomy, but he took it all with good humour:

> I hope to be excused for not informing other of my friends of the expected phenomenon, but most of them care little for trifles of this kind, preferring rather their hawks and hounds, to say no worse; and although England is not without votaries of astronomy, with some of whom I am acquainted, I was unable to convey to them the agreeable tidings, having myself had so little notice. If others, without being warned by me, have witnessed the transit, I shall not envy their good fortune, but gather rejoice, and congratulate them on their diligence. Nor will I withhold my praise from any who may hereafter confirm my observations by their own, or correct them by anything more.[14]

Horrocks knew that he had been given the opportunity to observe a very rare spectacle. He could measure the size and the orbital position of Venus with greater accuracy than had ever been achieved before. He also knew that the opportunity to take these measurements would not arise again for more than a hundred years. The Sun was the datum from which all the measurements of the planets were

Harrocks Hall. The hall stands above the Douglas valley between Parbold and Eccleston. It is the ancient seat of the Horrocks family.

Lower Lodge, Otterspool, from a watercolour by Edward Cox. The farm was a favoured candidate for Jeremiah Horrocks's birthplace, but it was demolished in 1862 and is now impossible to authenticate.

IRISH
EAST
SOVTH
Marton Mosse
Kirkham
Plumton
Westby hall
Newton
Clyston
Brining
Freckleton
Lethum
Warton
Lea hall
Penworthum
Langton chap
LAILAND
The Mose
Hesket
Laland
Howle
Keverdale
Law
Darwen flu
Walton hall
Houghton toure
Brundell
Clayton
Blackborn
Haudley
Hastingde
Darwen flu
Darwen Chap
Tockholes chap
Akrintan chap
Rossendale
Irwel
Entwisell
Hawcolm chap
Turton chap
Longworth
Walmesley Chap
Brandlessham
Egberden hill
Hall of the wood
Bradshaw
Ravenpik hill
Raventon
Smethels
Little Bolton
Harwich chap
Cockley Chap
Amsworth hall
Little Lever
Ratlif
Pilkin park
Werden
Charley
Asley
Yarrow flu
Hyll
Brotherton
Bankhall
Extonburgh
Eccleston
Charnok
Burgh
The pele
Tarleton wood
Holms park
The Lodge
Meales
Russord chap
Croston
Bispham
HUNDRED
Harrok hall
Aldington flu
Langne
Dowles flu
Andertonford
Blakerode
Worthington
Lastok
Dean
Snidle
Bolton
Great lever
Highfeild
Farnworth
SALFORD
Hulton park
Kingley
Ellynbrugh Cha
Wordley
Marton Mere
Skarbrik hall
Terlescowood
Marton
Parbot
Standish
Dowgles
Wrightinton
Standish hall
Wigan
Hay
Brinsap
Houghton Chappel
Harleston hall
Burscoo Abby
Halsall
Newhall
Lathom
Newbrugh
Ashurst
Dowles flu
Ince
Barton
New park
Ormiskirk
Croshall
Holland Chap
Pemerton
Shaker ley
Atherton
Worsley
Clifton
Alker chap
WEST DERBIE
Shelmarsdale
Taud flu
Orell
Winstonley
Bryn
Lowcon
Leigh
Tilsley
Boothes hall
Astley
Eckles
Formbye Chap
Aughton
Bickerstath
Lydiate
Cunscough
THE BRIGANTES
Barton
Altmouth
Ince
Mellen
Simons
Raynford chap.
Billing
Ashton chap.
Abram
Byram
Kilcheth
Morelees
Chatmosse
Whicklesw
Denaholm
Alt flu
Woode
HUNDRED
Little Crosby
Sefton
Forest
Mosbarrow
Ashton chap
Holcroft
Flyxton
Great Crosby
Kirkbye
Car
Windel
Bradley
Newchurch
Newton chap.
Letherland
Egleston
Irwel flu
Carinton
Partinton
Lerpoole Haven
Walton
Croxtath hall
Knowesley
Par
St Helins Chap
Wynwick
Sothworth hall
Rysley
Briche
Hollyn
Grene
Bankhall
West Darby
Prescot
Rixto
Warberton
ye black rock
Kirkdale
Hyton
Robye
Ranehill
Bewsey
Sankye
Wullston
Bold
Bollen flu
Wallasee
Earton
Lerpoole
Childwall
Farnworth
Thelwall
Lym
Wartre
Ditton
Appleton
Penketh
The Pele
Warrington
Toekseath park
Wotton
great
Agburg
Tarbok
Cuerdley
Gropnall
Garston
Hut
Hilbre
PARTE
Norton
OF

Above: John Speed's Map of Lancashire, 1610. This shows clearly the extent of Toxteth Park (Tockseath park) at about the time Horrocks was born. Other places mentioned in the script are Liverpool (Lerpoole), Hoole (Howle) and Harrocks Hall (Harrok hall).

Liverpool in 1680 by John Eyes. Hidden behind the castle (*right*) was the small harbour, the 'Liver Pool', and beyond this was Toxteth. Jeremiah's father was buried at St Nicholas Chapel (*left*). It still exists, but has been rebuilt several times.

Above and right: The Ancient Chapel of Toxteth. Although modified in 1774, much of the fabric is from the original 1618 building. It was well known to Horrocks in his youth, as was the first incumbent Richard Mather, who emigrated to New England in 1635.

The Dutch Wing, Emmanuel College Cambridge. The wing was built in the 1630s when Horrocks was an undergraduate student. It became known as the Old Court.

Nicholas Copernicus by Frederyk A. Lohrman. Copernicus's great achievement was to recognise that the motions of the planets could be greatly simplified by assuming that they all orbited around the Sun.

Tycho Brahe by an unknown artist. Brahe, the last and greatest of the naked-eye astronomers, studied and recorded the skies for twenty years. Kepler's work on planetary orbits would have been impossible without access to Brahe's data.

John Flamsteed by an unknown artist. In 1675 John Flamsteed became the first Astronomer Royal. His untiring support for the works of Horrocks and the recent publication of his papers have added much to our knowledge.

John Wallis by Gerard Soest. John Wallis was a friend of Horrocks during his student days at Cambridge. Wallis became a founder member of the Royal Society, and in the 1670s he was the editor of Horrocks's posthumous works.

St Michael and All Angels, Hoole. Built in 1628, it achieved the status of a full parish church in 1641. There is no evidence to support the theory that Horrocks was curate, but the church has a strong claim to Jeremiah as an important part of its history.

Carr House, Bretherton, built in 1613. The first-floor window has been suggested as the place where the observation of the Transit of Venus was made, but this was one of the best rooms in the house and the second-floor window seems more likely for the use of a lodger such as Horrocks.

The Founder of English Astronomy by Eyre Crowe. A romantic Victorian painting of Jeremiah Horrocks observing the Transit of Venus.

Crabtree Watching the Transit of Venus by Ford Madox Brown. Crabtree appears as an older man than his age of twenty-nine would imply, and the telescope is of a much later make than 1639. However, Brown has researched his subject and correctly concluded that William Crabtree's wife and children were present.

taken. When Venus was actually visible on the face of the Sun, her position could be found to a fraction of a minute of arc at her closest approach to the Earth. This compared to an accuracy of only about 1 minute of arc when the planet appeared as the morning star close to the Sun.

> If therefore on the other hand, we can observe her apparent place within a minute, it is clear that we shall ascertain her real longitude in her orbit within the third part of a minute; whereas when the planet is in other situations, a whole degree scarcely affects the apparent place of her longitude, especially in her greatest distances from the Sun, when observations of her are most frequently and correctly made; moreover both these and other observations plainly prove that the mean motion of Venus has never yet been determined by astronomers with sufficient accuracy. In the second place, no other observation shews so correctly the longitude of the node of Venus; for the telescope which I employed on this occasion is much more accurate than those generally used.[15]

When he came to write up his findings, Horrocks drew attention to the minute size of Venus. It appeared as only about a quarter of the size which previous astronomers had calculated when they were observing it as a bright object against a dark background, rather than a dark object against the Sun. Gassendi had already discovered a similar effect when he observed the transit of the planet Mercury. The results implied that the planets were either smaller than had been supposed or they were much further away. Horrocks suspected the latter, even though he knew it would further reduce the status of the Earth in the universe. It meant that the Earth could no longer be considered the largest of the planets, but was much smaller than Jupiter and Saturn and also possibly, according to his theory, smaller than Mars.

The transit of Venus had been observed and recorded for the first time in history. It was an excellent achievement. It was very fortunate for Horrocks that this rare event took place at the very time he was embarked upon his astronomical career. Even so, the observation on its own was little more than a footnote in the science of astronomy and any capable observer, knowing the time and day of the transit, could have achieved the same results as Horrocks. He deserved the recognition, however, because of his tireless observations of the planets, because of his refusal to accept tables which he knew to be wrong and because of his obsessional regard for detail. He was the only person able to calculate that the event would happen. Nobody else had predicted the transit. It was what Jeremiah did next with his findings, however, which showed him to have the hallmark of a genius.

CHAPTER EIGHT

The Parallax of the Sun

Every astronomer knows that Copernicus put the Sun at the centre of the known universe. But Copernicus had no idea about the scale of his model. He did not know the true size of the Sun or its distance from the Earth. This distance is obviously a key parameter to the scale of the universe, and it became known as the Astronomical Unit. In the 1630s, the Astronomical Unit had never been measured, but most leading astronomers had arrived at wildly inaccurate estimates. It was the ambition of Jeremiah Horrocks, like those before him, to find a value for this important distance.

The first attempts to measure the Sun's distance were made in the ancient world. Hipparchus, the greatest of the old-world astronomers, thought it was possible to calculate the distance from the measurement of a lunar eclipse, when the shadow of the Earth could be seen on the face of the Moon. He produced a diagram showing the angles which had to be measured. Hipparchus was held in such esteem that many people, including Kepler and Lansberg, tried to measure the solar system by his method, but they all seemed to misinterpret his diagram. Jeremiah Horrocks gave credit to Lansberg for his efforts, but he did not agree with the conclusions:

> Astronomy is much indebted to the learned and laborious Lansberg, as for other things, so chiefly for his explaining of the Diagramme of Hipparchus, which it is strang that so great masters of astronomy, such as Albategnius, Copernicus, Tycho Brahe, and Longomontanus should never be able to find out, but should make there hypothesis to disagree among them selves, as Lansberg demonstrates at large: Yet I doubt the Hypothesis of Lansberg himselfe, if they be brought to the tryall of this

> touchstone, will be found to be faulty, as well as the rest: but of this more hereafter: for the present it may be considered that he doth something wrong Kepler for he sayeth that Tycho Brahe, wth Christianus Longomontanus and Kepler, do make the semidiameter of the Sun 15' 0" and the Angle of ye halfe cone of ye Earths shadow 16' 2" which accussation is indeed true of Tycho and Longomontanus; but Kepler doth sufficiently cleare himselfe therof and sayth that the aforesaid angle is 14' 0" which exactly agrees with the diagram of Hipparchus, to wch Lansberg in another place confesseth that Kepler's hypothesis do answere.[1]

The diameter of the Sun, seen from the Earth, is about 30 seconds of arc, or half a degree. The Earth casts a shadow, a cone of darkness which reduces to a point in space where an observer could see the Earth just covering the Sun, exactly as the Moon covers the Sun during a total eclipse. This point is further away than the Earth from the Sun and the angular size of the Sun in the sky is therefore smaller. The difference between the two angles is the solar parallax, and when this is known, the distance to the Sun is a simple calculation. The only place where the Earth's shadow can be seen is on the surface of the Moon. Hipparchus knew that if he could estimate the size of the shadow cast on the Moon, then in theory he could find the apex of the shadow cone and hence the angle of the cone.

The theory is quite sound, but there are serious problems in practice. Firstly, the Moon is quite small compared to the Earth, so only a small fraction of the Earth's shadow falls on the Moon. This makes it very difficult to measure the diameter of the shadow circle accurately. Secondly, the difference in the angles is so very small, only a few seconds of arc, that the angles have to be measured to less than a second of arc to get any degree of accuracy. This was an impossible requirement in ancient times, and very difficult even in the twentieth century. In the ancient world, the Sun was thought to be much nearer to the Earth. The expected parallax was twenty times larger, but even this was almost impossible to measure.

In his 'Astronomical Exercises',[2] Horrocks systematically picks out the errors in Lansberg's work. The diagram had been misinterpreted and there were a number of corrections to the observed angles which

Lansberg had not applied. Horrocks concludes that the method could not be used to find the distance to the Sun:

> Thus have I plainly and truly shewed ye error that is Lansbergs hypothesis: The next thing is to shew how to amend it: a task (I must confesse) that I know not how to accomplish: it is an easy labour to espy faults, but how to redress them is not found with so much ease: but as Kepler sayth 'Inertia more est philosophia: vivamus nos et ex exeramur' I will therefore set upon this labour, and though I be not able to compasse my desire, yet it may be I may give some light for a further discovery. And first we must take leave of Kepler, to imagine those things for wch we cannot readily give a reason: for it is useful in Astronomys to preserve those embryous which have not a perfect & compleat forme, since the issues of the cr[e]at[o]r are like bears whelps, that receive their further shape by their dams licking of them.[3]

After the strange metaphor of the bear whelps, he goes on to tell us that he would like to spend more time studying the problem but does not feel he can afford the time:

> I would very gladly (if it would be) retaine the equall motions of the Sun and Moon, together wth the eccentricity, and other the hypotheses of Lansberg: and so fitte the shadows semidiameter together wth other things thereon depending; to his hypotheses, as that they would accurately answere all observations, without a new moulding of all, which is a labour too great to be performed by such stolen hours as I can afford to bestow upon this study. And I have great cause to hope that it may be done for I can not imagine that his tables answering so many observations of all ages, can in the main body be faulty; but only that they are out of joint in some small member, such as this of the shadows semidiameter: howsoever it be thus I attempt it.[4]

The layman would tend to think of the distance to the Sun as being expressed in millions of miles or kilometres, rather than as a parallax in seconds of arc. Astronomers, however, see everything in terms of angles on the celestial sphere and prefer to talk about distances in terms of parallax. This is because the distances are calculated by measuring very small angles.

Parallax is a familiar phenomenon. Everybody knows that the left eye has a slightly different view of the world from the right eye, and nearby objects are seen from a slightly different angle by each individual eye. The brain is very good at estimating distances through this binocular vision, using the parallax between the left and right eyes to turn the world into a three-dimensional picture. We can easily tell when we are looking at a three-dimensional object as opposed to a flat, two-dimensional image.

A child on a roundabout may observe that as he rides around on his fairground horse a nearby lamppost changes its position in relation to a row of poplar trees in the background. By observing at either end of the roundabout which individual trees line up with the lamppost, it would be possible, knowing the dimensions of the roundabout, to obtain an estimate of the distance to the lamppost. A more careful observer might notice that the poplar trees themselves change their position against the mountains on the horizon, and if the mountains were assumed to be at an infinite distance, then the distance of the poplar trees could also be calculated.

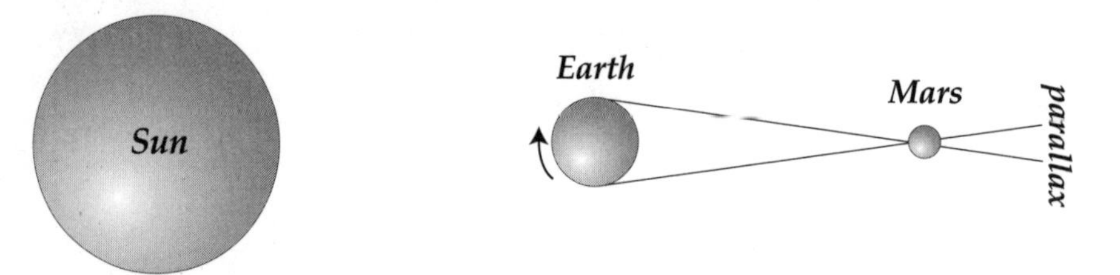

FIG. 7. The parallax of Mars

When the planet Mars is closest to the Earth, two observations can be made in the same night from points several thousand miles apart. This can be achieved simply from the rotation of the Earth. Mars is seen to move against the background of stars, but the movement is so small that the parallax could not be measured until the invention of the micrometer.

As we ride on the Earth around the Sun, it should be possible to see the nearer stars move against the distant stars in the background. In this case, the roundabout has become the Earth's orbit around the Sun. In the seventeenth century, nobody had ever managed to measure such a parallax. In fact, it was put forward as a cogent argument against the Copernican view of the universe. If the Earth was in motion, then the nearer stars should change their positions against the distant stars, so why was it that the star patterns remained fixed as the Earth in her orbit moved the vast distance around the Sun? The Copernicans argued that the stars must be at very great distances, too far away to detect the parallax, but they had no idea just how great the true distances were. The angles of the stellar parallax were so small that they were not measured until the nineteenth century, only a few years before the problem was greatly simplified by the application of photography.

One parallax had, in fact, already been detected – that of our nearest neighbour in space, the Moon. In ancient times, Hipparchus was able to make a good estimate of the distance to the Moon by comparing her position against the Sun as seen from two different points on the Earth. A partial eclipse was seen at Alexandria at the same time as a total eclipse of the Sun was seen at Rhodes. In the seventeenth century, astronomers wanted to find a similar method to measure the parallax of the Sun but this proved to be a very difficult problem. Kepler's laws gave a precise relationship between the periods of the planets and their mean distances from the Sun. This meant that if only one of the planetary distances could be measured, then all of them could be calculated. If the distance from the Earth to the Sun could be found, then the scale of the solar system and the size of the orbits of all the planets could also be found.

The horizontal parallax of the Sun offers a simple way to find the Astronomical Unit. As the Earth rotates on its axis, the Sun should appear to move against the distant stars behind it, and therefore all that is necessary is to measure the movement of the Sun against the background stars from dawn until sunset. The problem for the astronomer is that the Sun is not prepared to share the sky with the stars. The only time the stars are visible at the same time as the Sun is during a total eclipse, and it is impossible to have more than one eclipse in a day.

If Jeremiah Horrocks wanted to find the parallax of the Sun, he had to find another method of measuring it. He knew that with his observation of the transit of Venus he had the most accurate figure ever measured for the size of Venus seen from the Earth. He also had similar data for the planet Mercury with the results of Gassendi's observation. He knew that the angular diameters of the planets provided a clue to the scale of the solar system, and he began to wonder if he could use the results for Venus and Mercury to arrive at any new conclusions.

From his observation of the transit of Venus, Horrocks had made a direct measurement of the size of Venus viewed from the Earth. A simple calculation gave him the size of Venus seen from the Sun. He assumed for the sake of the argument that the distance from the Earth to the Sun was 98,409 units. This strange figure seems to be calculated from a mean distance of 100,000 units after allowing for the elliptical flattening of the Earth's orbit. The distance from Venus to the Sun was 72,000 of these units and the distance from the Earth to Venus was therefore 26,409 units. He had measured an angle of 1' 16", or 76 seconds of arc, for the angular diameter of Venus from the Earth. A simple calculation gave an angle of 28 seconds for the diameter of Venus as seen from the greater distance of the Sun.

Horrocks also knew that by using Gassendi's observation of the transit of Mercury, he could find the angular diameter of Mercury from the Sun. Gassendi had measured 20 seconds for the diameter of Mercury on the face of the Sun when seen from the Earth. Horrocks estimated the distance from Earth to Mercury as 67,525 units and the distance from the Sun to Mercury as 38,806 units. Another simple calculation gave the figure of 34 seconds of arc for the parallax of Mercury from the Sun. The figures for Mercury and Venus are far from equal, but Horrocks noted that they were both of the same order of magnitude. He knew that Gassendi's figure for Mercury was not very accurate. Mercury was a smaller planet than Venus and at a greater distance from the Earth. He was quite right in this assumption. While Gassendi had measured 20 seconds for the diameter of Mercury, the true figure was only about 11 seconds of arc, which would make the parallax of Mercury only 17 seconds from the Sun. Horrocks's own figure of 76 seconds for Venus was also an overestimate by about 25 per cent, but the orders of magnitude were correct.

He began to wonder whether the parallax of the Earth from the Sun was a similar magnitude. Having calculated the horizontal parallax of Venus and Mercury, he turned his attention to the planet Jupiter:

> Kepler supposes that Jupiter covers about fifty seconds by twilight. My proportion gives thirty-seven; the difference is not very great, and may be explained by Jupiter's brightness which increases his appearance. I have often compared Jupiter with Venus, which may be done with certainty, as they shine so equally. On the morning of the 25th February 1640, I thought him rather less; on the 2nd March, I thought him equal or perhaps rather larger; on the 6th, I thought him evidently larger. The diameter of Venus, at that time, was 0' 24", according to my estimate; and that of Jupiter about the same quantity. I do not suppose that this calculation is so accurate that a fault of a few seconds may not have arisen in it, either from the variable altitude of the planets, or from the degree of clearness of the diurnal light; but the conjecture is sufficiently satisfactory to my own mind, since it is clear that Jupiter does not differ perceptibly from the proportion of the other planets.[5]

This estimate was for the size of Jupiter when seen from the Earth. What Horrocks really wanted was the size when seen from the Sun, but knowing the positions of Earth and Jupiter in their orbits, it was a simple matter to calculate this angle. In the case of the outer planets, the variation throughout the year is not very great. The figure of 37 seconds fitted Horrocks's theory very well. He went on to study Saturn, the most distant of the known planets. When the rings of Saturn were edge-on to the Earth, they were invisible and the planet appeared as a normal sphere, but when the rings were viewed almost full-on, the planet seemed to be even larger than Jupiter. Kepler and an astronomer called Remus had estimated 30 seconds for the diameter. Jeremiah accepted this figure. Like the parallax of Jupiter, it fitted his theory well. He claimed a parallax of 30 seconds, but he changed his units from diameters ('by ye telescope') to semi-diameters in the same sentence:

> Therfore Saturn which is 10 times higher than ye Earth shall have a diameter 10 times greater than ye Earth, that so it may appeare to ye Sun

> as great as ye Earth doth, and Saturn by ye telescope is found to extend to 30", therfore his semidiameter is 15", eyther to ye Earth or Sun, for the difference is not much: therfore the Earth's parallax is also 15", which ye 60th part of the Sun's semidiameter…[6]

By '10 times higher than ye Earth', he means ten times the distance from the Sun. His conclusion, had it been true, would have been known to posterity as Horrocks's Law. He deduced that all the planets appeared the same size when seen from the Sun. All the parallax figures he had calculated implied an angular diameter of about 28 arc seconds, viewed from the Sun. Where the angles were a little away from this figure, the discrepancy could easily be attributed to errors in the observation. It followed that the parallax of the Earth from the Sun was also about 28 arc seconds. Unlike the other planets, the Earth

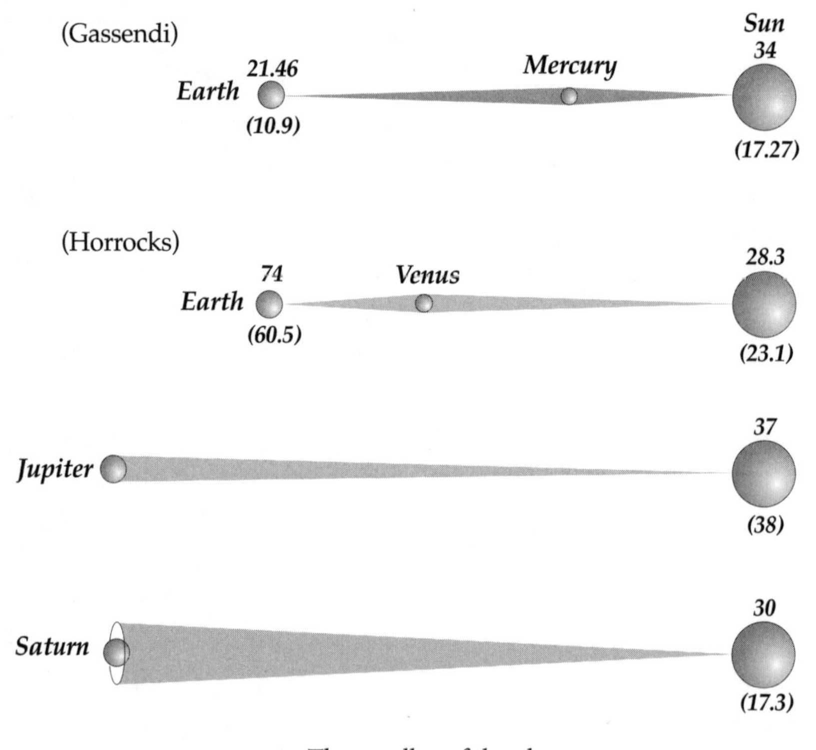

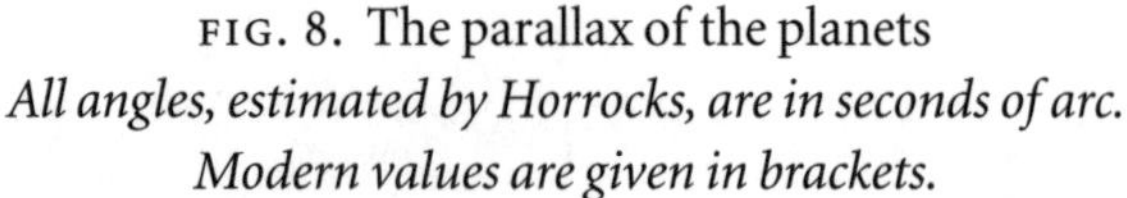

FIG. 8. The parallax of the planets
All angles, estimated by Horrocks, are in seconds of arc.
Modern values are given in brackets.

was the only one whose true size was reasonably well known. Using the figure of 28 arc seconds for the Sun's horizontal parallax, it was easy to calculate a distance of about 100 million kilometres for the Astronomical Unit. The modern accepted values are 17.4 arc seconds and 149 million kilometres. At first sight, Horrocks's estimate may not seem a very accurate figure, but it should be seen against the previous best value of 118 seconds (implying a distance of 22 million kilometres), calculated by Kepler. It then becomes obvious that Jeremiah Horrocks was the first astronomer to have any idea of the true scale of the solar system.

He knew that the planet Mars did not support his theory, but he explained this on the grounds of the reddish light which gave Mars a weaker glow than the other planets and made it seem smaller than it was. He cheated a little by pointing out that other astronomers, whose figures he had rejected for the other planets, had claimed a diameter of nearly two minutes for Mars at its closest conjunction.

> The planet Mars loses by comparison with the rest: and certainly does not exceed the assigned proportion. I suppose this is owing to his light being so remarkably obscure, for none of the planets sheds a feebler glow, or diffuses fewer rays. In the beginning of the month of March 1640, Mars appeared much less than Jupiter, thought they were in reality equal. He emits however a stronger ray by twilight when he is nearest to the Earth, and sometime appears so immensely large that he is mistaken for a new star: On this later occasion he seems nearly equal to two minutes, a quantity which perhaps he reaches; there is some doubt however on this point, in as much as no other planet, Jupiter and Venus not excepted, actually attains this dimension, though apparently they do not fall much short of it. But there is no need of hesitation when others extend the diameter to six or seven minutes . . .[7]

His observation of Venus had convinced Horrocks that the figures usually quoted for the angular size of the planets were far too high. His predecessors thought the planets were much larger than they actually were, compared to the size of the Sun, but that they were all much smaller than the Earth. Horrocks correctly deduced that the solar system was much larger than astronomers had previously thought, but this conjecture implied a second revolution, comparable with Copernicus's. Where Copernicus had calculated the Sun as 139 times the volume of the Earth, Horrocks calculated that the Sun was larger by a factor of 309,000. Furthermore, he calculated that Earth was not the largest planet, and correctly deduced that Jupiter and Saturn were many times larger. He wrongly deduced that Mars was also larger than the Earth, but he was not too happy about this conclusion.

The following table was compiled by Horrocks to compare his estimate of the solar parallax to those of other astronomers. He shows the solar parallax in terms of radius and not diameter and they are therefore only half of the values previously quoted. An extra column is included to show the distance in millions of kilometres and the bottom line is added to show the modern accepted values:

Astronomer	*Parallax of the Sun (Earth Radius/AU)*	*Distance (Earth Radii)*	*Astronomical Unit (millions of km)*	*Sun:Earth (Ratio of Volumes)*
Ptolemy	2' 50"	1210	7.7	166
Albertegus	3' 0"	1146	7.3	142
Copernicus	2' 55"	1179	7.5	162
Tycho Brahe	2' 54"	1183	7.5	139
Longomontanus	2' 35"	1334	8.4	196
Lansberg	2' 13"	1551	9.86	343
Kepler	0' 59"	3470	22.1	3,469
Horrocks	0' 14"	15,000	95.4	309,333
[Modern value	0' 8.5"	24,000	149	1,200,000][8]

Some have criticized Horrocks for his 'unscientific' calculation of the solar parallax. There was nothing unscientific about it. Everything was carefully reasoned. Later in the century, Halley and Newton estimated the distance of nearby stars such as Sirius and Canopus by assuming they were the same brightness as the Sun. They knew full well that it was very unlikely that all stars would have the same absolute brightness, but their reasoning did give the correct order of magnitude for the stellar distances. It showed that it would be almost impossible in their century to measure the distances of the stars by the method of parallax. Horrocks's calculation involved the parallel assumption that the planets would all look the same size when they were viewed from the Sun. If Horrocks's law did not hold for every planet, he knew that it would still give the right order of magnitude for the scale of the solar system. This was indeed the case. It meant that his estimate of the stellar distance was more than four times as large as the previous estimate, given by Kepler. And it meant that for the first time the planets were reduced to their correct, diminutive size against the great bulk of the Sun. Horrocks had envisaged a solar system far larger than Copernicus had done, but he knew that nobody would believe the incredibly large scale he had assigned.

Later in the century, when Hevelius published Horrocks's work, it seemed that the publisher had deliberately changed the parallax to give a distance of only ten million kilometres, presumably because he thought the paper would be ridiculed if he published the actual figure calculated by Horrocks. Even Newton, working fifty years later than Horrocks, used a smaller and less accurate figure for the astronomical unit because he did not believe the universe could be so large. 'I am not quite certain about the diameter of the Earth as seen from the Sun,' he wrote. 'I have assumed it to be 40 seconds, because of the observations of Kepler, Ricioli, and Vendelini do not allow of its being much greater. The observations of Horrox and Flamsteed

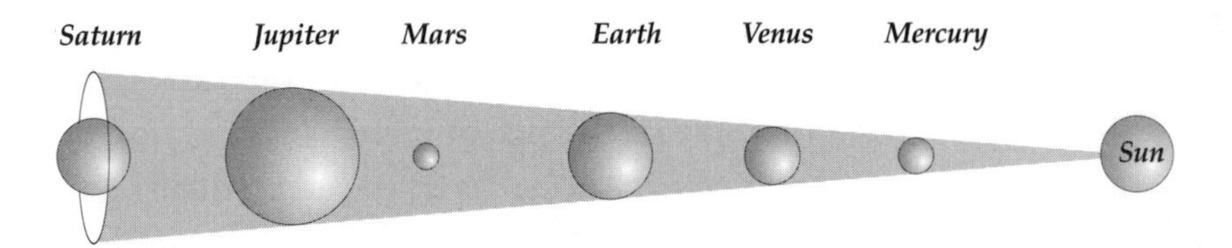

FIG. 9. Horrocks's Law

All planets are the same angular size when seen from the Sun (Mars was the exception that proved the rule, angle = 28 seconds). The 'law' was quite wrong, but gave the right order of magnitude for the scale of the solar system and Horrocks's estimate was far better than that of any astronomer before him.

make it somewhat less.' By the time the second edition of the *Principia* came out (1713), Newton had accepted that the figures proposed by his own countrymen were closer to the truth. It is an exaggeration to say that Horrocks was the first man to have the vision to appreciate the enormity of the universe as a whole. He had no idea of the existence of the furthest galaxies, but he took a great step into the unknown and he was the first to grasp the true scale of the solar system.

> Since therefore it is certain that the diameters of the five primary planets, in mean distance, appear from the Sun 0' 28", and that none of them deviate from this rule, tell me, ye followers of Copernicus, for I esteem not the opinions of others, tell me what prevents our fixing the diameter of the Earth at the same measurement, the parallax of the Sun being nearly 0' 14" at a distance, in round numbers, of 15000 of the Earth's semi-diameters? Certainly, if the Earth agrees with the rest as to motion, if the proportion of its orbit is to that of the rest be so exact, it is ridiculous to suppose that it should differ so remarkably in the proportion of its diameter. For it is incredible that of the six primary planets the diameter of one should be as much as 2', or as others make it 6', whilst all the rest should not exceed 0' 28". I have not within reach the opinions of other astronomers; but every one must believe what he sees for himself, and to me such a parallax seems absurd.[9]

Note that in the above passage Jeremiah Horrocks, revelling in his enthusiasm, confuses his readers by using both diameters and semi-diameters, which is why parallaxes of both 28 and 14 seconds appear in the first sentence. He goes on to say that he realizes the immense distance involved is unbelievable 'in as much as it exceeds, by ten times or more, the opinions hitherto received which so many excellent astronomers have geometrically demonstrated from their observations on eclipses'.

In order to test his hypothesis about the tiny angular diameters of the planets, Horrocks devised a simple device for measuring them. He took a piece of card and simply made a small circular hole in it with an iron needle of known diameter. He held the card close to his eye and looked at the planets through the small hole. He then moved the card further away, to try to find the point where the disc of the planet exactly filled the hole. It needed a steady hand, and he probably devised a simple stand on which to mount the card, with a string to measure the distance from card to eye:

> On the 7th Jan in the present year 1640, the Sun being risen and diminishing the rays of Venus by his own light, an iron needle whose diameter was 8 parts at a distance of 43000 covered the planet Venus; therefore diameter was 0' 38".
>
> On the 29th of January in the same year, a needle of 5 parts covered Venus at the distance of 38300; therefore the diameter was 0' 27"
>
> In these observations I looked through a small opening made with a fine needle in a piece of card; by which method alone, even on a dark night, the diameters of the planets appear to be wonderfully reduced; so that, unless you are very strong sighted, you can scarcely discover either planet or fixed stars which deceive the naked eye from their rays being so entirely cut of by the narrow opening.[10]

When Horrocks came to write up his account of the transit of Venus, he included an interesting paragraph, right at the end, stating clearly that he intended to write a full treatise on the parallax of the Sun. It is a tragedy that he never lived to complete this promised treatise. He claimed to have thought out new and easier methods of measuring the parallax of the Sun. He mentions the traditional method of finding the parallax, based on the diagram of Hipparchus and involving the measurement of the Earth's shadow during a lunar eclipse. Here we see the mature Horrocks, confident in his own ability and fully critical of many of those who had gone before him, even the greatest of the astronomers of the ancient world.

> I had intended to offer a more extended treatise on the Sun's parallax; but as the subject appears foreign to our present purpose, and cannot be dismissed with a few incomplete arguments, I prefer discussing it in a separate treatise, 'De syderum dimensione' which I have in hand. In this work, I examine the opinions and views of others; I fully explain the diagram of Hipparchus by which the Sun's parallax is usually demonstrated, and I subjoin sundry new speculations; I also shew that the

> hypotheses of no astronomer, Ptolemy not excepted, nor even Lansberg who boasts so loudly of his knowledge of this subject, answer to that diagram, but that Kepler alone properly understood it, I shew in fact that the hypotheses of all astronomers make the Sun's parallax either absolutely nothing, or so small that it is quite imperceptible, wheras they themselves, not understanding what they are about, come to an entirely opposite conclusion, a paradox of which Lansberg affords an apt illustration. Lastly I shew the insufficiency and uselessness of the common mode of demonstration from eclipses; I give many other certain and easy methods of proving the distance and magnitude of the Sun, and I do the same with regard to the rest of the planets, adducing several new observations.[11]

Galileo, when he turned his telescope on the stars, was not surprised to find that many more stars were visible than with the naked eye. What did surprise him was that, unlike the planets, he could see no more detail on the stars and he did not discover anything new about them. Even the brightest stars still appeared as points of light through the telescope. He calculated that they must be less than 5 seconds of arc in diameter. Did they shine by their own light or were they, like the planets, reflecting the light of the Sun? It seems a naïve question today, but it was not so in the 1630s. Jeremiah Horrocks also turned his attention to the size of the stars. Horrocks and Crabtree performed an experiment. They knew that when the dark limb of the Moon passed in front of one of the planets, the light from the planet faded over a measurable period of a few seconds before it disappeared. They reasoned correctly that the finite delay was a result of the finite area of the planet's disc. They performed the same experiment when the dark side of the Moon crossed in front of the stars in the constellation of the Pleiades. In this case, every star was extinguished in an instant when the Moon covered it. The implication was clear: the stars were points of light of an immeasurably small diameter.

Horrocks knew that the stars lay at an immense distance and he wanted to estimate how far away they were. It was widely thought that the stars appeared to rotate around the Sun once every 26,000 years. This was an illusion, due to what we now call the precession of the equinoxes. If there had been a planet which took this length of time to orbit the Sun, then it would be possible to calculate its distance using Kepler's relationship between period and distance. Other astronomers had already made this calculation. Lansberg, for example, calculated that the stars were at a distance of 280 million Earth radii. Horrocks, using the figures supplied by Lansberg, arrived at the over-precise distance of 131,411,357,378 English miles for the distance to the stars. This was, of course, a totally incorrect estimate and was based on invalid assumptions. He knew it was wrong and he tried to find other methods of finding the distance, but he also knew that it was impossible in his time to measure the small angles required by the method of parallax:

> All that can be said truly and certainly of ye distance of ye fixed stars is that wch Copernicus sayth that the Earths orbe is of an insensible bignesse in comparison of the spheare of ye fixed starres: for the parallax of the Earths orbe could not be perceived to be any by Tycho Brahes instruments. And therefore I am almost in opinion that their distance is greater than either of ye former opinions of Kepler and Lansberg; because they make a parallax that might perhaps bee observed by very great instruments; such as Tycho's were: especially considering that he could discerne 5" in them; as A. C. 1585 when he observed the height of Spica Virginis 25° 8' 55". And also Lansberg by his quadrants could discerne to the 1/6 part of a minute: it is like Tycho might have discerned a parallax of 8" or at least of 12", wch Keplers hypotheses allow; for though he make the distance of ye fixed starrs greater then Lansberg, yet, the Radius of the Earths orbe being also made greater, and that in a greater proportion, the parallax will be almost 12" greater a little then Lansberg. But such small things are not worth much disputation, and the neglect of this parallax of the fixed starres (if it were more then it is) could not breed any great error in Astronomy, which (alas) labours with many far greater imperfections.[12]

He went on to look for harmonic proportions and ratios between the size of the planets and their orbits. He reasoned that the Sun's diameter was the same size as the orbit of the Moon and he speculated that this proportion could be used to estimate the distance of the stars. A simple calculation, based on modern values, shows that this assumption was wrong, but it was difficult to prove in his time.

> The Moon in her apogaeum is distant 60 semidiameters of ye Earth, so that ye Moons orbe is as big as ye body of ye Sun: and both are as often containd in ye Earths orbe as ye body of ye Moon in hers. Also ye orb of ye Moon appears to ye Sun of ye same quantity that ye body of ye Moon doth to ye Earth: so that as ye Moons orb to ye Suns distance, so ye Moons body to her own distance; therfore ye Moons orb is a mean proportionall between ye Earths orb ye Moons body; and so ye Moon is contained in her own orb, as oft as her orb is contained in ye Earths orb: and lastly as oft as ye Earths orb contains ye Sun so oft the Moons orbe contains her. And for ye fixed stars it is like they make ye same parallax to ye Earths orbe, that ye Sun doth to ye Earths body, so that ye Earths orb is a mean proportionall between ye Earth and ye sphear of ye fixed stars; also ye Sun is a mean proportionall between ye Moon & ye sphear of ye Earth.[13]

In summary, we can say that in the year 1640 Jeremiah Horrocks became the first astronomer to calculate anywhere near the right order of magnitude for the distance from the Earth to the Sun. By the time his claim was published in *Venus in Sole Visa* (Danzig 1662), other astronomers, using more sophisticated equipment than that used by Horrocks, were coming to the same conclusion.

CHAPTER NINE

The Motion of the Moon

Johannes Kepler was the astronomer Jeremiah Horrocks admired the most among all his predecessors. When he came to write up his account of the Transit of Venus, his *Venus in Sole Visa*, Horrocks showed his enthusiasm by writing odes to the events and the people who moved him.

> I hasten therefore to that prince of astronomers, Kepler, to whose discoveries alone, all who understand the science will allow that we owe more than to those of any other person. I venerate with the greatest honour his sublime and enviable happy genius; and if necessary, I would defend with my best efforts the Uranian citadel of the noble hero who has so much surpassed his fellows, nor shall anyone while I live, violate his ashes with impunity. His death was an event that must ever have happened too soon; the science of astronomy received the lamentable intelligence whilst left in the hands of a few trifling professors who had kept themselves concealed like owls until the brightness of his Sun had set.[1]

We are left for a moment to guess who he means by the 'trifling professors' who concealed themselves like owls, but before naming them Horrocks goes on to describe his hero in more poetical strains:

> Who, mighty shade, shall sing thy praises? Who,
> Worthy so great a task, shall reach the stars?
> Who now shall chant thy fate? The modern seers
> Portend that heaven's disturbed by monsters which
> Are unintelligible to mankind;
> Perchance in pity thou dost still protect

The weaker minds of those whom thy decease
Hath robbed of nature's best interpreter.

He then describes the 'fictitious circles' of Ptolemy and his followers, and goes on to explain how each planet in its wandering through the sky followed precisely the path described by Kepler's laws:

Since such a guide is lost, what other now,
Deserving to succeed, can take the reins?
Or should the stars rebel, who can restore
Them to their course, and bind with closer ties
Their wandering ways? O! thou alone couldst take
The arduous guidance and shake the strong rein
To urge along the slothful retinue;
By thee restrained, the vulgar crowd
Dared not to climb the sacred car of heaven.
No devious course could cause thy thoughts to wander
In perplexity; fictitious circles
Could not enthrall thy loftier genius ;
But thy mind, intent on the sublime, with
Faithful hand traced the motions which the God
Of nature hath decreed. While yet the power
Was thine to guide their way, true to thy rules
Each planet in its ordered path revolved,
And all rejoiced to follow in thy train.

Horrocks feared that after Kepler's death the astronomical world would revert back to Ptolemy's cycles and epicycles. We now discover who the trifling professors were. Lansberg is referred to unkindly as 'the pompous Belgian'. Hortensius, the astronomer who had never accepted the work of Kepler and had even criticized the work of Tycho Brahe, is doomed for having insulted 'Pelides' dust'.

But now deprived of thee science declines,
Sinking in antiquated errors; all
The stars are buried as madness may devise,
And heaven's deformed by senseless violence!
Unhappy Germany! though torn by wars,

> The sword alone will not effect thy ruin;
> A heavier curse conspires to bring about
> Thy mind's destruction. 'Tis this encourages
> Hortensius to insult Pelides' dust;
> By this the pompous Belgian, bolder grown,
> Imposes on the world Perpetual Tables,
> And spurns the embers which a powerful flame
> Has sadly left; nor does he even fear
> Lest his bold thefts should haply be detected,
> Now that great Kepler's numbered with the dead.
> Chaos is come again, the world's unhinged,
> All things, in thee o'erpowered by fate, betray
> The noblest art to trifling sycophants.[2]

Kepler had made great strides with his laws of planetary motion. He was the first to discover the true nature of their orbits, but he never took the next logical step, which was to apply his theory to the motion of the Moon. It was very easy to claim by analogy that the Moon's orbit was an ellipse with the Earth at one focus, but Kepler never did. The reason was that he knew it was only a first approximation to the truth and that further corrections, known in the seventeenth century as equations, would have to be made to predict the true position of the Moon with any accuracy. Kepler did not attempt to solve this complex problem. Horrocks, on the other hand, perhaps with the brashness of youth, was only nineteen when he first applied his mind to the problem and in his 'Philosophical Exercises' he gives us some of his first thoughts. He is correct to assume that the Moon's motion around the Earth is analogous to the planet's motion around the Sun, but he is wrong to attribute the rotation of the Earth as the driving force:

> Ye Moons motion be caused by ye revolution of ye Earth and this revolution is lower in ye Aphelion than ye Perihelium as is most probable; it should seem that ye Moon's motion in like manner is slower in ye Earths Aphelion than in its Perihelium; and that in ye same proportion of inequality, that ye Earths diuranall revolution observeth, because it wholly depends theron. So that one revolution of ye Earth about its axis,

> whether it be performed in a shorter or longer time, shall yet have still ye same effect in moving ye Moon: for if it move slow & so be weaker yet it is longer about it; if fast, ye shortnes of ye continuance is recompenced wth the greater force of a fast motion. If this be so it will follow that ye physicall equation of time invented by Kepler, is to be omitted in ye Moon's motions, as being by their inequality recompensed. Though it be said that ye Sun doth help the Earth in moving ye Moon yet still shall be the same effect since that part of ye Suns help is in like sort unequally distributed.[3]

In the last sentence Horrocks clearly recognizes that the Sun, as well as the Earth, influences the motion of the Moon. This is a very far-reaching conclusion, but Horrocks cannot claim all the credit for it. Other astronomers before him had suggested that this was the case. The real problem was to quantify the Sun's influence and to calculate it mathematically. 'Besides ye Sun in helping the Earth to move ye Moon, doth not regard her latitude, but only distance in ye ecclipticke,' he wrote, though he followed this with second thoughts. 'Or may it not be that this is some cause of other inequality in ye Moon, than hath yet been taken notice of?' He believed that the true laws which governed the universe could be discovered by studying the motion of the Moon rather than that of the planets. He was getting closer to the principle of gravitation. He knew from his observations of Jupiter and Saturn that there were minor anomalies in the orbits, and he knew about the anomalies in the orbit of the Moon. He suspected that there might be more imperfections, still undiscovered, in the motions of the planets:

> Perhaps ye Moon, whose motions do in so many things imitate ye primary planets, epitomising their inequalitys & performing them in a shorter time, may bring to light those secular Equations, which Kepler esteems so certain. And therfore those varieties of inequalitys in ye Moon above ye rest of the planets, may be perhaps, but resemblances of those, wch ye planets cannot in so short a time discover.[4]

He wanted to find a terrestrial model for the Moon's motion. He took a plumb line, which he called a 'plummet', and studied the motion of the bob under different forces and conditions. This was not a simple pendulum experiment. The bob was suspended from a point, and he could make the bob perform circular and elliptical motions by experimenting with different starting velocities:

> Take a plummet and hang it by a thred, then remove it out of its place any way, & let it follow it's own course in ye ayr, & it will fall to the center & past it, till it stand stationary, & then return, still faster and faster till it be past the center again, & then ye motion will slacken, till it return back ye second time, and so will fall to & fro in a motion of libration just as the planets do in altum. But if you swing the plummet about with your hand it will describe an ovall figure a b c, and wch is worth noting the line a.b. or ye apsides (as I may call them) will move ye same way that ye plummet doth but much slower: so that if in one revolution ye furthest distance from ye center d, be in a, the next will be in c and this motion will be ye faster by how much the figure is more ovall, so that be not very long and have almost no roundnes, for then the apsides will hardly alter at all.[5]

This type of pendulum, sometimes called a spherical pendulum, is often quoted as a model for the planetary orbits. It is possible to make the bob trace out an ellipse as it revolves around the centre of its motion. It also satisfies the requirement that the bob moves slowly at greater distances from the centre and more quickly as it approaches the centre. The analogy breaks down, however, on closer study. The planets, unlike the plummet, are not drawn to the central point of their elliptical orbits, but to the focus, where the Sun is located. If the orbit is a circle, then the velocity remains constant and the analogy holds good, but in general the Sun is offset from the centre by an amount which depends on the eccentricity of the ellipse. The other area where the analogy breaks down is the strength of the force on the bob. In the case of a planet, the force is weaker when it is further from the Sun. In the case of the plummet, the force becomes stronger the further the bob is from the centre. Horrocks tried to allow for the eccentricity by introducing a shear force, performing the experiment with a steady wind blowing from one side. This had the required effect of moving the orbit off-centre. The model still suffered from

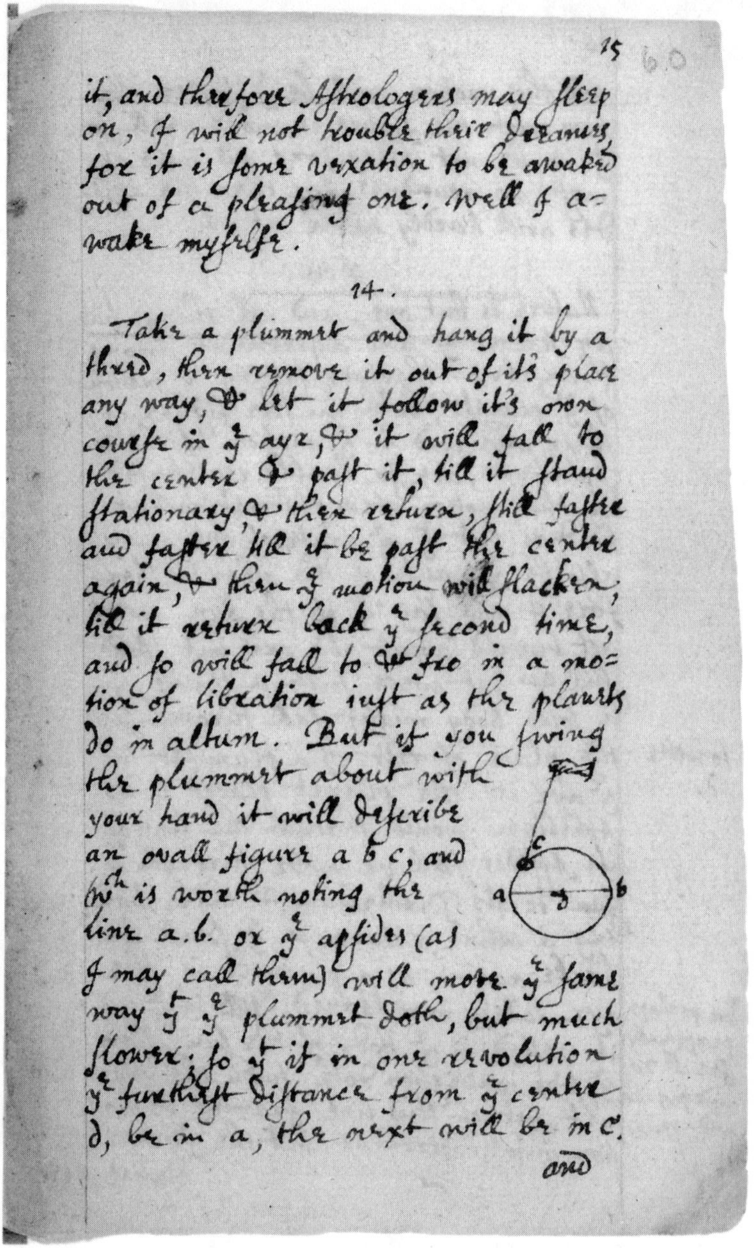

15 6.0

it, and therefore Astrologers may sleep
on, I will not trouble their dreames,
for it is some vexation to be awaked
out of a pleasing one. Well I a-
wake myselfe.

14.

Take a plummet and hang it by a
thred, then remove it out of it's place
any way, & let it follow it's own
course in ye ayre, & it will fall to
the center & past it, till it stand
stationary & then returne, still faster
and faster till it be past the centre
again, & then ye motion will slacken,
till it returne back ye second time,
and so will fall to & fro in a mo=
tion of libration iust as the planets
do in altum. But if you swing
the plummet about with
your hand it will describe
an ovall figure a b c, and
wch is worth noting the
line a.b. or ye apsides (as
I may call them) will move ye same
way yt ye plummet doth, but much
slower: so yt if in one revolution
ye furthest distance from ye centre
d, be in a, the next will be in c.
and

c
a d b

FIG. 10. Sheet from 'Philosophical Exercises'
Page from Horrocks's notebooks, using the analogy of the pendulum to explain the motions of the planets.

other defects, but he was able to deduce that the effect of the Sun on the Moon's orbit was to cause it to rotate or precess.

In his letter to Crabtree dated 1 July 1637, he speaks of 'attacking the Moon next as this had produced the greatest problem'. It was not long before he deduced that the first approximation to the orbit was an ellipse with the Earth at one focus.

One of the authorities who gave Jeremiah Horrocks credit for being the first astronomer to discover that the Moon's orbit about the Earth was an ellipse was Isaac Newton in the *Principia* (1686). Newton states that '*Horroccius noster lunam in ellipsi circum terram, in ejus umbilico inferiore constitutum, revolvi primus statuit*.' ['our countryman Horrocks showed that the Moon's orbit is an ellipse around the Earth, with the Earth situated in the lower focus'].[6] This statement seems to be no more than Kepler's law applied to the Earth/Moon system as opposed to the Sun/Planet system. Why was it, therefore, that other astronomers – Kepler in particular – had failed to grasp something which could be guessed so easily? The reason is that merely stating that the orbit is an ellipse is of no value in itself unless the claimant goes on to prove his case by determining the constants of the ellipse. The Earth, being situated near the centre of the orbit, is not the ideal place from which to determine the shape of the Moon's orbit. Horrocks was a brilliant observational astronomer, however, as well as a theoretician, and he was able to measure the constants of the orbit.

The complex motion of the Moon has generated a whole vocabulary of its own. The variations from the circular motion are known as 'inequalities'. There are many of these and in the ancient world several of them were known to Hipparchus and Ptolemy. The mathematical corrections are known as 'equalities' or 'equations'. The 'perigee' and the 'apogee' are the furthest and nearest approach of the Moon in her orbit. The 'sysygies' and 'quadratures' are the quarter points of the Moon in her orbit, corresponding roughly to the full and new moon and the two half moons. The 'eccentricity of the orbit' is a measure of the amount of flattening of the ellipse from a true circle. The 'libration' is the apparent oscillation of the Moon but, like the word parallax, it is loosely applied to any form of oscillation.

'Evection' was a term introduced after the death of Horrocks, but it describes the variation of eccentricity which had been detected by Hipparchus but which Horrocks was the first to formulate mathematically.

One of the main reasons for the study of the Moon's motion was so that astronomers could predict an eclipse of the Sun, and consequently the motion of the Moon had been studied in depth for thousands of years. Horrocks noted that there was a lot of inconsistency in the observations of the eclipses. When the same eclipse was observed from different places, for example, the times given were not always consistent with each other. Even different observers in the same place gave different times for the beginning and end of the same eclipse, an illustration of how difficult it was to determine accurate times before the pendulum clock became available. He also knew there was a variation due to the refraction of light through the Earth's atmosphere and this was not always taken into account.[7]

One of the observations which led Horrocks to his lunar theory was the phenomenon known as libration. It was well known that the Moon always presented the same face to the Earth as she moved around her orbit. The Moon rotates once on her axis for every orbit she makes around the Earth – the period of the orbit and the period of the rotation are identical. This means that in the time of Horrocks and for more than three centuries after him, nobody had ever seen the far side of the Moon. It was widely thought, however, that the Moon wobbled a little in her orbit, showing a few degrees of her hidden face at certain points of her path around the Sun. How could this wobble be explained?

First, consider a similar, but not identical, situation which arises with the Earth's orbit. In the northern hemisphere, the shortest day is 21 December, but for a few days afterwards the Sun rises even later. The sunset, however, is also later, so the 21st is still the shortest day. The longest day exhibits a similar anomaly. The longest day is 21 June. Sunset on the 24th is about a minute later, but sunrise is two minutes later, so the 21st is still the longest day. If the orbit of the Earth were a perfect circle, then the shortest day would have the latest sunrise and also the earliest sunset. Similarly, the longest day would have the earliest sunrise and the latest sunset. The reason why this is not the case is that the orbit of the Earth is not a circle. It is an ellipse. Consequently the time of sunrise and sunset are not symmetrical, they are skewed.

Now consider the Moon again. The Moon cannot 'wobble' as she makes her orbit. Jeremiah Horrocks was born too early to know anything about the conservation of angular momentum, but with his marvellous instinct for astronomy, he knew that the Moon rotated with the same constant angular velocity throughout her orbit. The wobble is an illusion created by the fact that on the Earth we are not situated at the centre of the Moon's orbit, but some distance away, at one focus of an ellipse. This gives us a skewed view of the Moon as it moves around us. We sometimes see it from a slight angle and this appears to us as a wobble. This effect became known as libration, and

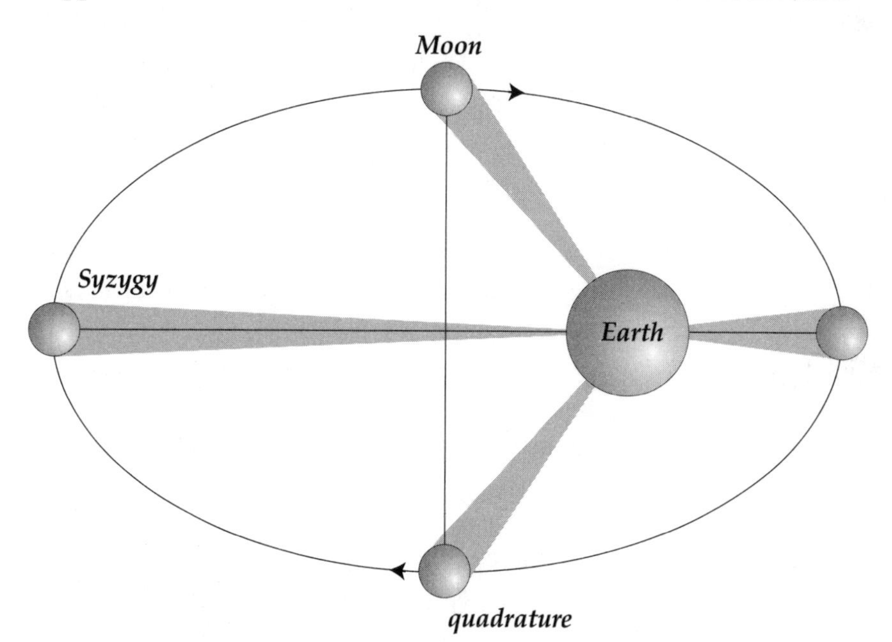

FIG. 11. Libration of the Moon

The Earth is not situated at the centre of the Moon's orbit. This is one reason why we are able to see part of the hidden surface of the Moon at some points in her orbit. Horrocks was able to use this effect to calculate the constants of the orbit.

Horrocks was the first to realize that libration could be explained by assuming an elliptical orbit for the Moon.

When Horrocks first began to measure the motion of the Moon in 1638, it did not take him long to deduce that the orbit was an ellipse, but he recognized that this was only a first approximation and that there were other factors which had to be determined. He discovered that the long axis of the ellipse, called the line of apses, rotated in a cyclic fashion, and he correctly deduced that the reason for the rotation was related to the position of the Sun. He suggested two corrections to allow for the perturbing effect of the Sun. One was to cause the orbit to rotate or precess in a cyclic manner and the second was to vary the eccentricity, or the amount of flattening, of the ellipse. In the terminology of the times, these corrections were called 'equations', in the sense that they helped to equate the predicted position of the Moon with the observed position. Horrocks put a lot of his short working life into his lunar theory. It was not until 1640 that he had perfected his second version.

Some idea of the degree of difficulty can be gained from the comments of other astronomers. When Edmund Halley urged Isaac Newton to complete his theory of the Moon and to publish it, Newton replied that the problem kept him awake at nights. In the last decade of the seventeenth century, when Newton was still trying to solve the same problem, he admitted that the problem was so difficult that it made his head ache. We might ask why was it that one of the few mathematical problems which Isaac Newton was unable to solve had already been solved by a boy barely out of his teens over half a century before him? To answer the question, we must first admit that Horrocks's solution was not the final answer. He provided three crucial steps in the right direction, but other 'equations' were required to perfect it. The second fact is that Newton quite rightly wanted to solve the problem from his Law of Gravitation. He could calculate very precisely the gravitational forces of the Sun and the Earth on the Moon at any point in her orbit. He could set up the equations of motion, but what he could not do was solve them. The problem is known as the three-body problem. The three bodies were the Sun, the Earth and the Moon, each with its own mass attracting

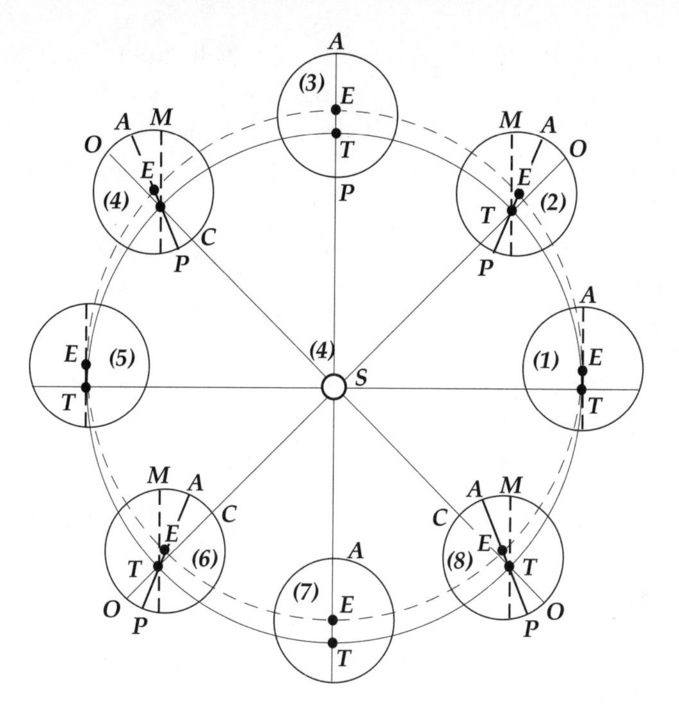

FIG. 12. Crabtree's diagram of Horrocks's lunar theory
The Earth is the point 'T' moving around the Sun 'S'. The small circle is the Moon's orbit and is actually an ellipse. The axis 'AETP' of the ellipse oscillates throughout the year.

the other two. It is a problem which is now known to be impossible to solve completely by analytical methods. Horrocks, on the other hand, was following the tradition of Kepler and looking for an algorithm to predict the position of the Moon at any time in the future. He did not know Newton's Law of Gravitation. He knew that the Sun was the disturbing factor and he applied corrections according to the position of the Sun.

In the third book of the *Principia*, Isaac Newton wrote that

> there are many inequalities in the Moon's motion not yet noticed by astronomers. They are all deducible from our principles [of gravitation and mechanics], and are known to have a real existence in the heavens. This may be seen in the hypothesis of Horrocks which is the most ingenious, and if I do not decieve myself, the most accurate of all[8]

Newton also described the lunar theory of Jeremiah Horrocks in a manuscript entitled 'A Theory of the Moon', in which he appears to give credit to Edmund Halley for suggesting the next correction beyond those suggested by Horrocks. Newton's passage is heavy-going for the layman, but it is of great value to the memory of Horrocks to quote the precise words of Isaac Newton, showing that Newton was aware of the genius of his predecessor and fellow countryman:

> The old astronomers tried to exhibit the motions of the planets by means of eccentric circles in solid spheres. By observations of the comets Tycho discovered that the heavens are not solid, nevertheless he retained circular motions, as he paid very little heed to the causes of the motions. At length Kepler first discovered from astronomical observations that the motions of the planets take place in ellipses round the Sun. Our countryman Horrocks, in order to avoid any inconsistency in nature, would have it that the Moon too revolves round the Earth in an ellipse. Just as Kepler showed that the planets describe areas proportional to the time by a radius drawn [from the planet] to the Sun, situated in the common lower focus of the orbits, so too, Horrocks supposed the Moon to describe, by a radius drawn from it to the Earth, situated in the lower focus of her orbit, an area proportional to the time; the eccentricity of the orbit, however, does not remain the same as is the case with the primary planets, but increases and diminishes in turn according to the position of the Moon's apogee with regard to the Sun. Halley, too, inferred from his astronomical observations that the upper focus of the Moon's orbit is borne in a uniform motion on the circumference of a circle whose radius is half the difference between the greatest and the least eccentricity, and whose centre revolves round the Earth with uniform motion at a distance equal to the mean eccentricity. In our previous discussions, we have attempted, by applying the theory of gravity to celestial phenomena, not only to investigate the causes of the motions, but also to compute the magnitude of these motions from their causes. And we supplied some examples of this in the motions of the Moon also. For it follows from this that the revolutions of all the planets arc ruled by gravity, and that not only are solid spheres to be resolved into a fluid

> medium, but even this medium is to be rejected lest it hinder or disturb the celestial motions that depend upon gravity.[9]

Whilst Horrocks was working on his lunar theory, William Crabtree had not been idle. Jeremiah Horrocks was not Crabtree's only scientific correspondent. Crabtree had become acquainted with Christopher Towneley of Towneley Hall, near Burnley. Towneley was a wealthy landowner, very interested in science and antiquity. He was also a practising Roman Catholic, and this meant that there were religious differences between himself and the Puritanical family of Jeremiah Horrocks. The men did not seem to allow their religion to come between them, however. When Crabtree explained the works of Horrocks to Towneley, the latter was quickly able to appreciate them. Crabtree also had a third correspondent, across the Pennines in Yorkshire. His name was William Gascoigne and he lived at Middleton, near Leeds. Gascoigne saw himself as more of a practical instrument-maker than an observational astronomer. He was the first to realize that the way to develop the telescope was by means of micrometer sights, a set of fine wires guided by a finely machined screw thread which would allow angles to be measured with far greater accuracy. He put a lot of effort into developing the micrometer and he was anxious to share his inventions with others. Crabtree wanted to introduce Jeremiah Horrocks to William Gascoigne. In August 1640, Crabtree wrote to Gascoigne, suggesting that they should arrange a meeting and offering to bring his friend and 'second self' Jeremiah Horrocks across the Pennines into Yorkshire. He knew how much Horrocks would appreciate Gascoigne's micrometer and he realized what a great asset it would have been had it been available to observe the transit of Venus. Another point of interest in their correspondence is the reference to the work of Galileo:

> In the mean time let me encourage you to proceed I your noble optical speculations. I do believe there are as rare inventions as Galileo's telescope yet undiscovered. My living in a place void of apt materials for that purpose makes me almost ignorant in those subjects: only what I have from reason, or the reading of Kepler's Astronomia Optica, and Galileo. If you impart unto us any of your optical secrets, we shall be thankful

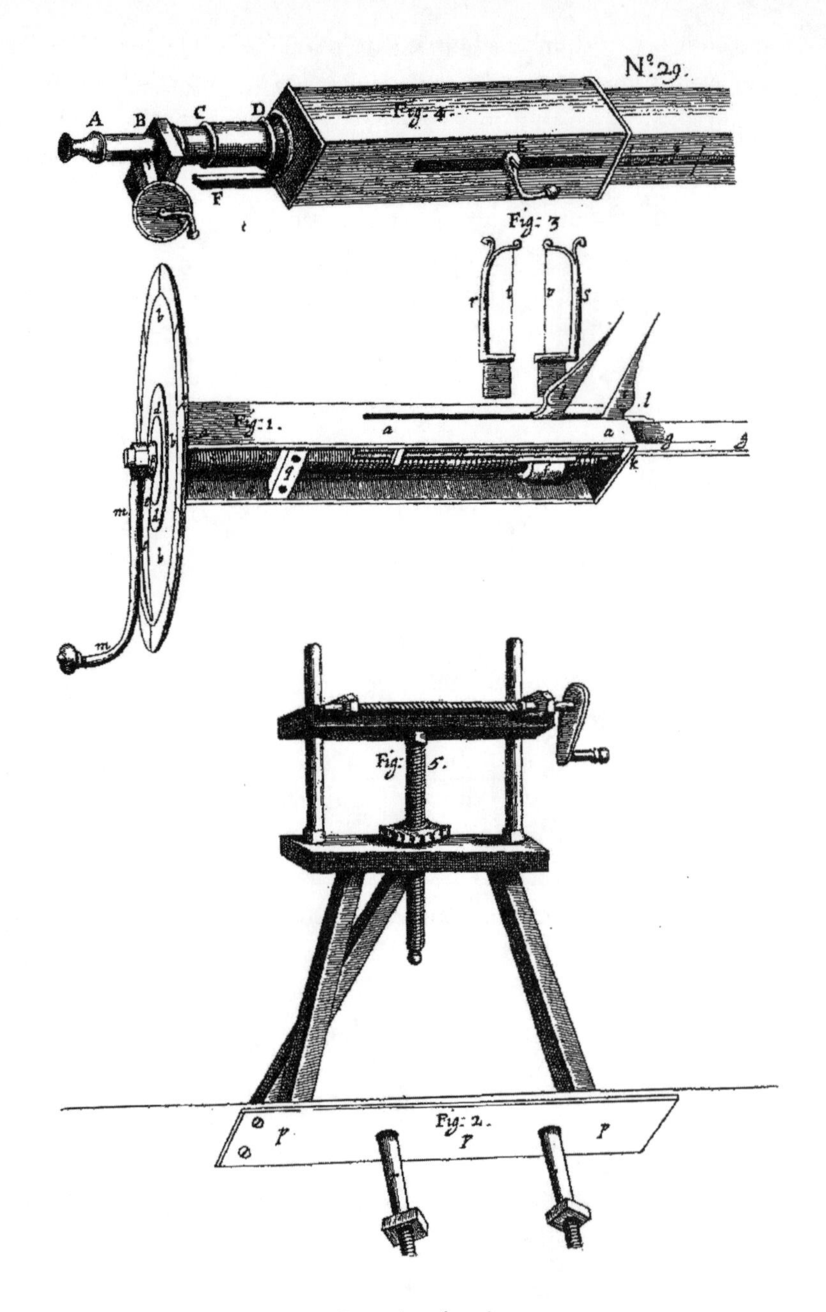

FIG. 13. Gascoigne's micrometer
William Gascoigne's micrometer (by Richard Towneley, drawn for the Royal Society in 1667).

> and obliged to you, and ready to requite you in anything we can. It is true which you say, that I found Venus' diameter much less than any theory extant made it. Kepler came nearest, yet makes her diameter five times too much. Tycho, Lansberg, and the ancients about ten times greater than it should be. So also do they differ as widely in the time of the conjunction. By Lansberg the conjunction should have been 16h. 31m. before we observed it: by Tycho and Longomontanus 1 day 8h. 25m. before: by Kepler, who is still the nearest to the truth, 9h 46m before. So that had not our own observations and study taught us a better theory than any of these, we had never attended at that for that rare spectacle. You shall have the observation of it when we see you. The clouds deprived me of part of the observation, but my friend and second self Mr. Jeremiah Horrocks, living near Preston, observed it clearly from the time of its coming into the Sun, till the Sun's setting; and both our observations agreed, both in the time, and diameter most precisely. If I can, I will bring him along with Mr. Townley and myself to see Yorkshire and you.[10]

There were other observations which William Crabtree wanted to discuss with Gascoigne. On 22 May 1639, an eclipse of the Sun was visible in England. Horrocks and Crabtree had both observed the eclipse and so, too, had Samuel Foster at Gresham College in London. Foster was quick to appreciate the merits of the northern astronomers and he had also become a regular correspondent with the group. As Crabtree started one letter to William Gascoigne, he could not resist having another carp at poor Lansberg on whom Crabtree and Horrocks had wasted so much of their valuable time:

> You shall also have my observations of the Sun's last eclipse here at Broughton, Mr. Horrocks's between Liverpool and Preston, and Mr. Foster's in London. Lansberg on eclipses, especially the Moon, comes often nearer the truth than Kepler, yet: it is by packing together errors; his diameters of the Sun and Moon being false and his variation of the shadow being quite repugnant to geometrical demonstration. His circular hypotheses, Mr. Horrocks, before I could persuade him, assayed a long time with indefatigable pains and study to correct and amend; changing and turning them every way, still amazed and amused with

> those lofty titles of perpetuity and perfection so impudently imposed upon them; until we found, by comparing observations in several places of the orbes, that his hypotheses would never agree with the heavens for all times, as he confidently boasts; no, nor scarce for any one whole year together, alter the equal motion, prosthaphaeresis, and eccentricity howsoever you will. Kepler's ecliptick is undoubtedly the way which the planets describe in their motions; and if you have read his commentary 'de motu Veneris' and his 'Epitome Astronomiae Copernicae,' I doubt not you will say his theory is the most rational, demonstrative, harmonious, simple, and natural that is yet thought of, or I suppose can be; all those superfluous fictions being rejected, by him, which others are forced so absurdly to introduce; and although in some respects his tables be deficient, yet being once corrected by due observations, they hold true in the rest, which is that argument of truth which Lansberg's and all others want.[11]

It is here that we discover that Jeremiah was still in touch with his cousin Thomas Horrocks, whom he had known at Cambridge and who had emigrated to America with his uncle (-in-law) John Cotton. Jeremiah asked his American cousin to estimate as accurately as he could the local time from the first recovery of the Moon's light after the lunar eclipse of June 1638 until sunrise on the same day. Thomas measured 65 minutes by the clock. By using Lansberg's tables, Jeremiah Horrocks calculated that the American observation had been made from a point 5 hours and 30 minutes (82.5 degrees) west of Goes in Holland. This was long before the Greenwich meridian had been accepted as the standard datum. The longitude was in error by about 28 minutes, or about seven degrees. The observation is traditionally thought to have been made from the place now known as Quidnick on Rhode Island. If this is the case, then the latitude was also in error by about 53 minutes, all of which combines to make a very poor case for the first determination of a longitude in the New World. Horrocks knew all about the uncertainty of the method. 'Do not rely on this,' he wrote characteristically, 'for the calculation is uncertain and the observation not absolutely exact.' The principle was perfectly correct, however, and later generations of astronomers recognized that the observation of an eclipse gave an excellent method for determining longitude.[12]

The correspondence sometimes turned to other matters. One topic which came under discussion was that of decimalization. The astronomers were not so much concerned with the coinage as with their angular measurements. They knew that a decimal system would do away with the need for clumsy fractions in scientific work. One of the worst bugbears was the measurement of angles. Angular measurement would be so much simpler if the circle were divided into a large number of decimal parts, each part smaller than one second of arc. This would do away with the sexagesimal system of degrees, minutes and seconds inherited from the Babylonians. Crabtree's letter proves that at least a few people in England were proposing a decimal system before it was adopted by the French, but they were, as usual, far ahead of their time. It was not surprising that they were ignored by the scientific establishment, in so far as it existed before the time of the Royal Society. Gascoigne suggested dividing the circle into a hundred million parts. Crabtree liked the idea and explained that he had already produced a decimalized version of Kepler's tables:

> Your conceit of turning the circle into 100,000,000 parts were an excellent one, if it had been set on foot when astronomy was first invented. Mr. Horrocks and I have often conferred about it. But in respect that all astronomy is already in a quite different form, and the tediousness of reducing the tables of sines, tangents, and all other things we should have occasion to use into that form; as also the inconveniences which we foresaw would follow in the composing of the tables of celestial motions, together with the greatness of the innovation, deterred us from the conceit. Only we intend to use the centesmes, and millesmes of degrees, because of the ease in calculation. I have turned the Rudolphine tables into degrees and millesmes, and altered them into a far more concise, ready, and easy form, than they are done by Kepler. My occasions force me to put an abrupt end to my unpolished lines, and without more compliments, to tell you plainly, but sincerely, I am your loving friend, (though de facie ignotus)
>
> William Crabtree[13]

Thus, by 1640 Christopher Towneley and William Gascoigne had entered the circle of the north country astronomical renaissance.

Jeremiah Horrocks was back at Toxteth, making measurements of the tides in the River Mersey. He knew that the Moon had an effect on the tides:

> The spring tides are always farr greater when ye Moon at conjunction and opposition is in ye equinoctial, then when shee is in ye tropicks. May not the Earth be stronger at those times in moving her, as well as shee in moving ye waters?[14]

It was well known that the tides were affected by the motion of the Moon, but Horrocks wanted to try to detect a pattern which would help him understand how the Moon could have this strange influence over the seas. On 3 October, Horrocks wrote to Crabtree to say that he had been measuring the tides.

> ye heavens remain free from any such impediment as shall, without miracle, eyther fully overthrow, or notably disorder that frame which nature hath appointed; as we see that ye motion of the sea, though subject to many casuall and physicall intentions and remissions, yet in its course is constant and unchanged, yea and the inequalitys of ebbing & flowing, of spring & neap tides are so regular as that they may be certainly foretold, yet still some little incertainty is left to ye dispose of contingent causes. So may not God hath given ye stars such laws, as though they do very neer obey, yet some small discrepancy meerly contingent and irregular should be reserved for ye only foresight of ye creator, that so man might not boast of any perfection, but by a limited skill be taught modestly to esteem of his own confined witte.[15]

Similar ideas appear in the *Opera Posthuma*:

> [The motion of the seas] ... has indicated many rare things to me ... It is strongly regular but is a subject to many variations of motion and remarkable inequalities, not previously noticed. The observations so far have continued for three months. However I hope that if I remain here for a whole year I may discover many secrets which may openly prove the motion of the Earth[16]

The parallel with Newton surfaces again. Horrocks seemed to know instinctively the right course to follow. He claimed to have discovered some very interesting patterns connected with the phases of the Moon, but he chose to wait until he had more observations and could draw some definite conclusions before developing his ideas and passing them on to others.

At the age of twenty-two, Jeremiah Horrocks had reached his prime. His ideas on gravitation and celestial dynamics were developing rapidly. He had determined the scale of the solar system. He had developed his system for the motion of the Moon. Soon, his work on the tides would add to the knowledge of the universe. His circle of scientific correspondents was opening up rapidly. William Gascoigne's micrometer would allow him to measure the positions of the planets even more accurately than before. A new revolution in astronomy was just around the corner. There was much for the two friends to discuss with each other. They arranged a meeting at Broughton early in the New Year, on 4 January 1641.

> A definite time will have to be fixed for me to visit you. If you can fit in the 4th of January it will not be a problem for me. I think I will be free then. Unless something out of the ordinary gets in the way, you can expect me.[17]

They had become close friends. Horrocks's high opinion of Crabtree was expressed in his description of the latter's observation of Venus on the face of the Sun. Crabtree's admiration for his younger friend is expressed in his letter to William Gascoigne. On the day appointed, Crabtree waited eagerly for his friend and fellow astronomer to arrive. He waited and waited. But Jeremiah Horrocks never came. It was a few days before Crabtree found out the reason why Horrocks did not keep his appointment.

Jeremiah Horrocks was dead.

CHAPTER TEN

War

William Crabtree wept. He took up the last letter he had received from Jeremiah. He put it with his other letters from his friend and he wrote down his feelings:

> Letters from Mr Jeremiah Horrocks to me dating from the years 1638, 1639, 1640 up to the day of his very sudden death on the morning of 3rd January, the day before he had decided to come and see me. Thus God sets an end to all earthly things. Ah departed friends (alas, the sadness of it all). O Horrocks most dear to me! Ah the bitter tears this has caused! What an incalculable loss![1]

Crabtree determined to keep the letters and to treasure them for the rest of his life. He wrote to William Gascoigne and he spent a whole page of his letter lamenting the death of Jeremiah Horrocks. The letter does not survive, which is unfortunate, because it probably mentioned the symptoms and cause of death. In Horrocks's last letters to Crabtree there is evidence that he may not have been well. He seems to realize that he has much to do and very little time left in which to do it. The reasons why he returned to Toxteth from Hoole are not clear, but the move could have been for reasons of ill health. On 12 December, he wrote to Crabtree that '... if I were not kept by a great necessity, by which I am either unwillingly detained at home, or compelled to journeys less pleasing, I would long since have hastened to you at Broughton, that I might more fully know what new matters you are giving your mind to.'[2] Horrocks complained that he had not made any astronomical observations for three months, but he thought that if he could have another year at Toxteth he would find out many things. His work on the tides was one example. His letter uses the wording 'As for yourself, continue your observations, and I will prepare to enter into them again as soon as I have settled my business (*mea negotia*)'. The phrase implies that it was his business affairs rather than his health that was keeping him from his beloved astronomy.

The average life expectancy in Horrocks's time was very short, due mainly to the very high rate of infant mortality and to the terrible epidemics which visited every community at least once in every generation. Horrocks's death was obviously not an infant mortality and there is no evidence of plague at the time. The most likely cause of sudden death for one so young would be a heart problem, but this is only suggested as a possibility because of the lack of any evidence of other causes.

The death of a close friend or relative is very hard to bear, but when the deceased is elderly there is a small consolation in that they have lived a normal life span and that death is part of the natural sequence of events. There is nothing harder to bear than the sudden death of a young person, and there is nobody who feels the bereavement more than the parents of the deceased. On the back of a letter from Crabtree to Gascoigne, a short note was added by John Flamsteed, the future Astronomer Royal: 'Mr Horrox father died May 3 1641: griefe for his sonnes hastening his own death'.[3] James Horrocks, the Toxteth watchmaker, knew the merits of his brilliant son as well as anybody. The implication is that James grieved so much at his son's death that he survived Jeremiah by only four months. Flamsteed's note is confirmed by the parish records of St Nicholas in Liverpool, where the burial of 'James Harrocks, watchmaker' on 1 May 1641 is recorded in the bishop's transcripts.

There is nothing to indicate what happened to poor Mary Horrocks, Jeremiah's mother. At some point, her son Jonas left for Ireland, taking with him some of his dead brother's papers. Jonas eventually returned home and his name appears in the Liverpool municipal records a few years later. The Horrocks family soon found that they were not alone in their bereavements. It was no consolation to them, but within the next few years practically every family in the land lost at least one relative. During the 1640s, fathers, husbands and sons were killed in the greatest conflict ever fought on British soil.

The rift between Charles Stuart and Parliament had been moving towards a crisis point for a long time. Charles I became king of England and Scotland in 1625. He believed in the divine right of kings and thought he should be allowed to rule his country without the help of Parliament. The rift between the king and the country was not simply political. It was complicated by religious differences which divided England, Scotland and Ireland. Charles supported the Church of England in resisting the demands of the Puritans. The communion table, for example, became a major point of issue. The Church of England wanted the table to stand in the east of the church, with an altar rail to separate it from the congregation. The Puritans preferred to have the table in the nave, where the congregation would often use it to place their hats and belongings during the church service. When Charles insisted on his changes, the Puritans thought he was trying to move closer to the Roman Catholic faith. It was hardly an issue on which to go to war, but it generated a lot of contention. In the 1630s, this and other issues helped to fuel the migration to the American colonies.

In Scotland, the Presbyterian church had made great strides in recent decades and the movement was strong enough to propose that the Presbyterian code be accepted as the official religion of Scotland. Charles refused to acknowledge the Presbyterians and thereby engendered more enemies of the Crown. The Parliament which met in 1628 provided the last chance to set up a peaceful reform for the government of the country, but in his capacity as the ruling sovereign Charles had the power to dissolve Parliament and this is precisely what he did. Charles did not call a single Parliament during the whole of the 1630s. He ruled without it for eleven years, but he found to his cost that he did not have the power to levy taxes without the consent of Parliament. One of the few ways for him to raise funds was through ship money. He therefore made frequent demands for the money and ordered all towns, inland as well as coastal, to pay the tax. The result was even more discontent.

The conflict which became known as the English Civil War was a misnomer. It was in fact a British Civil War, with the Scots and Irish just as involved as the English. The Welsh were also involved to a lesser extent, and there was plenty of action in the border counties. In 1637 there was a demonstration at St Giles's Church in Edinburgh, where women pelted the clergy and a mob tried to lynch the bishop. When the situation in Scotland deteriorated, Charles recalled Thomas Wentworth from his administration in Ireland, created him the earl of Strafford and gave him the job of putting down the revolt in Scotland. It was the Scottish revolt that forced Charles to recall Parliament, as he needed money to pay the troops to put it down. The Short Parliament, as it became known, lasted only three weeks. Most of the Commons, to the king's dismay, sympathized with the Scottish grievances, and they were not prepared to go to war over the issue. Charles dissolved the Parliament and began to seek another solution to the problem. Later in 1640, Parliament reassembled itself in what became known as the Long Parliament and one of the first items on the agenda was to pass a law calling for regular Parliaments every three years.

The king wanted Strafford to lead the Irish Army into Scotland to quell the revolt and as a result Strafford came under attack from both English and Scots. The earl of Strafford became the scapegoat. The leader of the Commons, John Pym, impeached Strafford so that he could not take his seat in Parliament. Pym introduced a bill of attainder against Strafford, a summary condemnation to death by special Act of Parliament. The bill was passed by a large majority and Strafford was executed. Charles might have been able to save Strafford, but he was intimidated by popular rioting and he dared not intervene.

The Civil War began in earnest in 1642, when Charles raised his standard at Nottingham. It was not the first time that civil war had broken out in England, but no previous conflict had involved so many people, no war had ever divided the country so sharply and no war had ever caused so many deaths on English soil. The king gained support from northern and western counties and from part of the midlands, but the eastern and southern counties sided mostly with Parliament. It was significant that Charles did not have the support of London, from where most of the revenue to fight the war was to come. Family sided against family and friends sided against friends as the conflict escalated out of control.

In October, the first great battle was fought at Edge Hill in Warwickshire. It was not a conclusive battle, but it showed that the Royalist cavalry, under the dashing command of Prince Rupert of the Rhine, was superior to the Roundhead cavalry. The king's army advanced as far as Brentford, but it was obvious that they did not have the resources to take London. Charles beat a retreat and set up his headquarters in Oxford. In the following year, 1643, the Royalists seemed to gain the advantage. In May, Sir Ralph Hopton with his Cornish army won a great victory for the Royalists at Stratton. In July, Hopton's army was joined by that of Prince Rupert and the combined forces successfully took the important city and port of Bristol. This gave the Royalists control over much of the West Country, but they were unable to take Gloucester, where the Roundheads remained firmly in command. The first battle of Newbury was fought in September 1643. Again, neither side gained a clear victory, but on the fourth day the Royalists ran out of gunpowder. If it had not been for the lack of gunpowder, the course of the war might have been quite different.

The county of Lancashire was divided in its loyalties. At the outbreak of the war, the aristocratic families of Stanley and Molyneux supported King Charles. Liverpool consequently declared for the king, but the gentry and the trading community had very different loyalties and the town was as sharply divided as the rest of the country. In 1643, as the Royalists made gains in the West Country, it was the Parliamentarians who gained the upper hand in Lancashire, and in the same year as the port of Bristol was taken by the Royalists, the port of Liverpool was taken by the Roundheads. Hastily built mud walls were constructed, with fortified gates, to hold the town against the Royalists. The castle, which had not been used as a fortress for many years, was restructured, repaired and strengthened to hold a garrison. Lord Stanley, the earl of Derby, escaped to the Isle of Man to avoid the worst of the fighting. His seat at Lathom was besieged by Parliamentary forces, but it held out for several months under the leadership of his gallant French wife, Charlotte de Tremouille.

In 1644 Prince Rupert headed north with the main Royalist army and by May of that year he had Liverpool under siege. The garrison held out for several weeks until 13 June, when Rupert's forces broke through the defences and launched a night attack. There followed the bloodiest night in the history of Liverpool. There was hand-to-hand fighting in the depths of the night, with no quarter given. It was total chaos, with most people not knowing who was friend or foe in the darkness. When dawn broke on 14 June, 160 bodies lay dead in the streets. The Royalists plundered the countryside for miles around. The tradition is that James Horrocks's house in Toxteth was looted and many of Jeremiah Horrocks's papers were burned.

After taking Liverpool, Prince Rupert relieved the countess of Derby, who had been under siege for four months at Lathom Hall, near Ormskirk. Twenty thousand loyalists flocked to his banner and he moved his army north and east to cross the Pennines into Yorkshire. By July, Rupert was outside the walls of York. There followed the bloody and decisive battle of Marston Moor on 2 July 1644. It was the beginning of the end of the Royalist cause. Not only Rupert's troops but also many Roundhead soldiers fell at Marston Moor. William Gascoigne cared only for the development of his micrometer, but he took a Royalist commission and fought for the king. He survived Marston Moor, but died early the following year at a minor skirmish near Melton Mowbray.[4] Other accounts claim that William Crabtree also fought at Marston Moor. If this is true, he survived the battle but not for very long. He was dead before the end of the month and was buried at Manchester Collegiate Church on 1 August 1644.

The Towneley family also suffered bereavement. There is a moving story about Colonel Charles Towneley, who died at Marston Moor. When Mary Towneley, his wife, heard that he had fallen in battle, she ventured across the Pennines to York in the hope that at least she would find the body of her husband. There were four thousand bodies on the battlefield, many badly mutilated and all stripped of their clothing and possessions. It was a hopeless task. One of the

Roundhead officers, seeing a Royalist lady searching the battlefield and surrounded by Roundhead troops, approached her. He was sympathetic, but advised her that she was in great danger from marauding soldiers and that she should leave for home as soon as possible. He then arranged for a trustworthy Roundhead soldier to carry her to safety on his horse. As they left the battlefield, Mary Towneley asked the soldier if he knew the name of the officer who had helped her. 'Oliver Cromwell,' was the reply.[5]

All three of the young astronomers were dead and the flowering of the astronomical renaissance in the north of England was over. The parallax of the Sun was a hundred million miles away. The Civil War was the brutal reality.

Cromwell learned a great deal from the battle of Marston Moor and he began to train a newly modelled professional army. The New Model Army became known as Cromwell's 'Ironsides'. When the antagonists met again at Naseby in 1645, the Ironsides carried all before them and the result was another decisive victory for Parliament. The commander in chief at Naseby was Thomas Fairfax, with Cromwell as second in command, but when the war was over it was Cromwell, as the leader of the Commons, who gained control of the country. The Royalist cause was lost but there was no obvious way forward. The army had to be retained until there was a stable government and the country was at peace again. The question in everybody's mind was where would Cromwell turn from here?

There was a powerful section of the army who knew exactly what they wanted for the way forward. They were called the Levellers and they proposed that Britain should become a republic, a country that could be governed by the people without the need for a monarchy. They also suggested the popular idea that everybody should be free to worship as they wished. Cromwell, however, although he himself could not see the next step forward, would not be swayed by their logic. Some of the Levellers mutinied. They were court-martialled and they lost their cause. A second movement started up, calling themselves the True Levellers, who advocated a system of communism whereby all property belonged to the state. The True Levellers even went as far as to set up a commune, where they planted their oats and turnips. They became known as the Diggers. The True Levellers took their radical ideas too far. Landowners strongly resented the idea of losing their property to the state, and the new levelling ideas never gained popular support as an acceptable form of government.

The fighting should have been over, but Charles somehow managed to escape from the Isle of Wight and find his way to Scotland. By promising the Scots that he would allow Presbyterianism to become the official religion of Scotland, the very point on which he had refused them eight years earlier, he managed to raise another army. A second civil war followed but it did not escalate very far. The Scots were defeated in a three-day battle near Preston. Cromwell then found himself with no choice but to put the king on trial. It was a foregone conclusion that King Charles would be found guilty of treason to his own country, and the sentence for treason was death. Charles coolly declared that as a ruling monarch he could not be put on trial and that the proceedings were illegal. When the fateful day of execution arrived, Charles I's last hour was his finest. As his severed head was displayed outside the palace in Whitehall, there were no cheers, only a great sigh and a gasp from the assembled multitude.

The interregnum brought a degree of stability. Cromwell was firmly in control as long as his army remained loyal to him, but he could still not find an acceptable way forward. He refused to become king, but in 1653 he declared himself to be Lord Protector. He ran the country with great efficiency until his death in 1658, but there was one occasion when he went so far as to dismiss Parliament, as Charles had done before him. This was not what the people had fought for, and when Cromwell died there was only one direction for the country to turn. The restoration of the monarchy was the only way forward. Charles II, who hid in an oak tree to escape from the Roundheads after the battle of Worcester, was brought home from exile in France.

During the Civil War, the universities of Oxford and Cambridge found themselves on opposite sides of the conflict. Oxford became the king's headquarters and was obliged to give him support, even though many of the dons were Roundhead sympathizers. Cambridge, by contrast, was firmly in the centre of the Cromwell country. When the Puritans gained the ascendancy, Horrocks's old college, Emmanuel, came into great favour because of its Puritanical foundation and policies. The claim that Emmanuel produced more masters than any other college at this time is hardly surprising when it was seen as the most Puritanical college in the university. Sir Humphrey Mildmay, grandson of Sir Walter Mildmay, the college founder, supported the king in the great conflict: a classic example of division within a family, where individual loyalties were the opposite of what might be expected.

Under Cromwell, the Puritans had gained control of the whole country and they achieved all they had fought for. What they did not gain was the permanent support of the people. The religious fervour, the monotony of prayers and psalms at all times, was not conducive to a majority who did not want their lives to be dominated by religion. The banning of the maypole, the closure of the theatres and the general killjoy attitude towards dancing and other forms of recreation became their downfall. In 1647 there was a rebellion against the Puritans in Canterbury when they tried to suppress Christmas festivities on the grounds that they were pagan. The English people wanted to enjoy themselves after so many years of war and the Puritan religion made no allowance for them to do so.

The great political and theological struggles which marked the middle years of the seventeenth century meant that progress in the sciences was pushed into the background, but in spite of this, some significant progress was made. Gresham College, founded at the end of the sixteenth century, managed to survive all the upheavals. After Cromwell's death in 1658, the buildings were taken over as a barracks, but at the Restoration it returned to its former function of an educational establishment.

In the 1640s, during the Civil War, a small group of enthusiasts began to meet in London to discuss matters of natural philosophy. The group consisted mainly of academics from Gresham College and became known as the 'invisible college' because it had members but no buildings or material possessions. Early in the 1650s, two of the group's most active members, John Wallis and John Wilkins, were given new posts at Oxford to replace some of the unfortunate academics who had supported the losing side during the Civil War. These two were the prime movers in setting up the Oxford Philosophical Society. The Oxford group included men like the young Christopher Wren, who went on to become a professor of mathematics and astronomy, and Robert Hooke, who became curator of the Royal Society.

Soon after the Restoration of the monarchy, the initiative moved back to London, where a philosophical society was formed in 1660 with the consent of Charles II. The king showed a great interest in the proceedings of the society and in 1662 he granted a charter allowing it to call itself the 'Royal Society of London for the promotion of Natural Knowledge'. A second charter was granted in the following year. The Royal Society was granted a coat of arms and acquired a grand silver mace weighing 150 ounces after the fashion of Cromwell's 'bauble', the parliamentary mace. Thus, in 1662, when Horrocks's *Venus in Sole Visa* was published in Danzig, the Royal Society was still a new body. A great deal had happened since the author of the treatise had died in 1641. It was a generation later and things had moved forward in the world.

> While yet on Earth, the youthful pastor trod,
> He read the word and traced the works of God;
> The courses of the stars prophetic saw,
> Unwound their order, and defined their law.
> And yet a loftier view his eye could scan
> For this lost world salvation's glorious plan
> The firmament of souls redeemed from night,
> The centre Jesus, and the circle light.
> A Sage's love, a young Apostle's zeal,
> The head to reason, and the heart to feel
> With truth and mercy graced the preacher's tongue,
> And o'er his life a holy radiance flung.
> That meteor – life, soon lost to vision here,

Now shines unclouded in a glorious sphere;
Yet here its light his bright example gives,
And here in fame undying Horrox lives.

Revd A. B. Whatton[6]

CHAPTER ELEVEN
Opera Posthuma

The death of Horrocks was not quite the end of the astronomical renaissance in the north of England. William Crabtree found Gascoigne a good correspondent and he described some of Horrocks's works to him, in particular his theory of the lunar motion. Crabtree's circle also included the brothers Christopher and Charles Towneley of Carre Hall and Towneley Hall respectively. Charles Towneley died at Marston Moor, Crabtree died less than a month later and Gascoigne died in the same year. It was therefore the battle of Marston Moor and its aftermath which brought about the end of the northern astronomical renaissance. After Marston Moor, the only member of the group to survive was Christopher Towneley, and he was an elderly man who had been a sponsor and a co-ordinator of the group rather than an active participant. There was just one small piece of good fortune. Richard Towneley, son of Charles and nephew of Christopher, was also a dedicated scientist. The two surviving Towneleys knew that if they did not collect together the remaining works of Horrocks, Crabtree and Gascoigne, then all would be lost to posterity.

Thus the manuscripts came to Carre Hall. It was there that Christopher Towneley employed another astronomer, another Jeremiah, to work on the manuscripts of the deceased Horrocks. His name was Jeremiah Shakerley. He is often taken to be a Lancashire man by virtue of his contact with the Towneleys, but in fact Shakerley claimed to have been born near Halifax in Yorkshire. It was Shakerley who, thirteen years before Hevelius, became the first astronomer to bring Horrocks's works to the notice of the public, but unlike Hevelius, he did not publish a complete treatise. He produced three works in all, *Anatomy of Urania Practice* (1649), *Almanac* (1651) and *Tabulae Britannicae* (1653). In all three he made good use of Horrocks's manuscripts and he generously acknowledged the work of his predecessor.

Shakerley was a competent astronomer, but he differed from Horrocks in several respects. Principally, he was interested in the more lucrative study of astrology. His publications concerned the ephemerides of the planets, the constants needed for astrologers to calculate the positions of the planets at any date and time, and he knew that the figures he had obtained from Horrocks were the most accurate available at the time. Shakerley's reputation is a little tarnished by his continual appeals for sponsorship: he obviously did not have the independent means to support himself and his research. Although Christopher Towneley gave him board and lodgings, this was not enough for him and he hoped to gain more lucrative financial rewards for his efforts. Most of our information about Shakerley comes from his correspondence with the astrologer William Lilly of London. The latter was a horoscope-caster of great repute. He was the official astrologer for Parliament during the Civil War and had prospered from his almanacs and horoscopes.[1]

This is a good place to reflect on Horrocks's supposed poverty. There is a complete absence in Horrocks's correspondence of any form of begging or appeals for sponsorship. Poverty is a relative term. Horrocks was a sizar at Cambridge and he was not seen as wealthy by the standards of the times, but he was lucky enough to have a hard-working and supportive family who were prepared to help him through university and to finance his researches. Shakerley did not enjoy the same good fortune, but he did have one asset which Horrocks did not. He was prepared to travel to any lengths to pursue his astronomy and when he discovered that a transit of Mercury was due to take place in 1651 he was determined to observe it. He was so determined that, when he found the transit would not be visible in Europe, he travelled all the way to India to make the observation. Thus, whereas Horrocks hardly left his native county, Shakerley was prepared to travel the world, in spite of his poverty, to advance his astronomy. He, too, died young and his working life of five years was hardly longer than that of Horrocks. Towards the end of his life, Shakerley changed his attitude towards astrology, not to disbelief but to disillusionment with the dishonesty of some of its practitioners. The upheavals of the Civil War were also a great stumbling block for Shakerley: although his works were published, they were not popular enough in times of war to inform the country of the merits of the man who had preceded him by about ten years.

In the 1660s John Flamsteed, in a letter to John Collins at the Royal Society, seems to challenge the date of Horrocks's death and adds some gossip he had picked up regarding the deaths of Crabtree and Gascoigne. The letter is quoted here with some reservations, because nearly all the dates are wrong, but it is of some value because it shows the source of the false dates. The information was no more than idle gossip which Flamsteed heard from a Mr Wroe, the warden of Manchester, which he recorded in good faith. Wroe happened to be one of Crabtree's neighbours:

> I thinke tis not anywaies evident from Crabtrees letter, that his freind Mr Horrox died not the 3d of Jan: 1640. Except I have a misse translated something, pray peruse it agane and let mee know. For besides the note on the back of the letters I find that in a letter of Crabtrees to Gascoigne dated March 18 1640/41 hee much laments the death of Horrox: Mr Gascoigne was slaien in our Wars, I beleive in the yeare 1642 for I find no letters either of his or Crabtrees to him after that of Jun 21 1642 if I have not miswritten. Crabtree lived much longer, I believe till 1652 if his Neighbour Mr Wroe informed mee truely I shall tomorrow write to Mr Townly, and will make one part of my desires to be ascerteined of the time of his death. Present my services to Mr Oldenburge and tell him I shall be carefull to make enquirys of the old people and how they have lived as hee desired but haveing no great acquaintance I feare I cannot procure any long list. I waited last night for the Suns Eclipse but it was such cloudy and rainy weather I could never see him plainely after 5 aclock
>
> I am Sir Ever Yours
> John Flamsteed[2]

The errors quoted in the letter include Horrocks's death, which, strictly speaking, should read 3 January 1640/1. Gascoigne's death occurred in July 1644, not 1642. Crabtree did not live until 1652. He died in February 1645. Flamsteed was a very meticulous man and the errors are out of character, but he acted in good faith and he was not to know that the information he had been given was in error.

By September 1664, John Wallis had gathered together all he could find of the remains of Horrocks's work. His book, the *Opera Posthuma* – actually published under a far longer title which was guaranteed to confuse the reader but was very much in the fashion of the times – was almost ready for the press. The publication was a lengthy process and there were many delays which slowed things down. Money was one factor. The Royal Society was, as usual, slightly embarrassed for funds. Then came the terrible plague of 1665, by far the worst pestilence in living memory. The plague was followed by the Great Fire of 1666, when London became the scene of total chaos. Conditions for publication in a London full of bereavements and lost fortunes could hardly have been worse. Some of Horrocks's papers were lodged with the bookseller Nathaniel Brooks, whose premises were destroyed by the fire. It is not known which of his papers were lost, but they may have been the originals of the letters written by Horrocks to Crabtree. It was very fortunate that John Wallis had completed most of his task by that time, so that although the originals had been lost in the fire, the letters had been copied by Wallis and were ready for publication. Whereas the original letters were written in English, Wallis had translated them into Latin for the benefit of the international community. They can be translated back into English, but something is lost by two translations and Horrocks's original wording can only be guessed.

After the terrible havoc of the Great Fire, the Royal Society was forced to vacate its premises at Gresham College. The building was needed to house the Royal Exchange. It took several years for things to return to anything like normal. The outcome was that it was not until the 1670s that the Royal Society was again ready to consider publishing the *Opera Posthuma*. This was ten years after *Venus in Sole Visa* had been published and thirty years after Horrocks had died, but in the intervening years many of the ideas in the posthumous work had leaked out and as the publication date came nearer, a lot of interest was generated.

John Flamsteed managed to obtain a copy of Horrocks's original treatise on the transit of Venus and with his meticulous thoroughness he discovered several small errors in the Danzig publication. The diameter of the Sun as used by Horrocks was in error by about 20 seconds, or about 1 per cent. This was a minor point and made little difference to the conclusions. What puzzled Flamsteed was that Hevelius quoted Horrocks's estimate of the Sun's parallax as 40 seconds when Horrocks had clearly calculated only 28. The reason may have been to give the paper more credence, because nobody would believe the figure quoted by Horrocks, but Flamsteed attributed the error to a misreading by Hevelius of one of Horrocks's diagrams. He wanted the account of *Venus in Sole Visa* to be included in *Opera Posthuma*, but he was overruled. The transit of Venus had already been published and was available to researchers. The Royal Society thought there was no need to publish it a second time. Flamsteed also knew from his enquiries that there was still more to be discovered about Horrocks and he was determined to find all he could. In December 1670 he wrote to John Collins, expressing concern about the delay in the publication of the *Opera Posthuma* and offering to help with the editing. Flamsteed was interested in Horrocks's observations of Saturn. He was aware that Saturn and Jupiter exerted some influence on each other. He rightly thought the effect might be of some importance, and he knew that with the elapse of over twenty years the errors detected by Horrocks would be larger and easier to measure:

> Sir
>
> You once I remembred proferred to lend me Mr Horrox'es papers which are now in your hands. If Sir You have not procured them yet to be printed nor any bookseller offers to undertake them I would accept of your offer gladly for I Have libe[rty] now to peruse any author and I would gladly try, if that deviation which is now very perceptible in the motion of Saturn was not in his time sensible. For if I remember aright one of his books conteines the observed distances of the planets from the

> fixed stars with his other observations from which nothing as yet hath been deduced[3]

The aspect which interested John Flamsteed most of all was the theory of the Moon. The Royal Greenwich Observatory was yet to be founded, but it was generally recognized that an accurate theory of the motion of the Moon was the key to the important problem of finding the longitude at sea. At this time there were three rival theories of the Moon's motion. The first was by Ismael Bulliadus (*Astronomia Philolaica*, Paris 1645), the second by Thomas Streete (*Astronomia Carolina*, London 1661) and the third by Vincent Wing (*Astronomia Brittania*, London 1669). Flamsteed first discovered Horrocks's work on the Moon in the letter to Crabtree dated December 1638, but after a careful study he found it was not as accurate as the theories of the three later astronomers which he had also studied. In 1672 he wrote that 'I can not but much approve of the forme of Mr Horrox theory, and having considered it severall times I find wee might almost as easily compute her place in her orbe as [we calculate] the elliptical place of any other planet'. It was a great disappointment, but what Flamsteed wrote was correct. All Horrocks had done was to fit an ellipse to the orbit of the Moon. The ellipse was a significant step in 1638, but the result was of little use in predicting the Moon's position with the required degree of accuracy. For a time, it looked like the demise of Jeremiah Horrocks's lunar theory.

Soon afterwards, however, Flamsteed wrote again to John Collins. This time, he was elated. It seems that his previous discovery was no more than Horrocks's first thoughts on the theory of the Moon. Horrocks had spent a further two years thinking about the problem and trying to find ways to improve it. Horrocks's fully developed theory was explained in the letter from Crabtree to Gascoigne dated 1642, more than a year after Horrocks's death. This was a very fortunate survival and it underlines the skill of Crabtree as a mathematician, for if Crabtree had not interpreted the work of Horrocks correctly, he could not have written it up accurately for men like Gascoigne and Flamsteed to use. Flamsteed tested the new theory with some figures of his own, and this time he found excellent agreement between Horrocks's later theory and his own observations of the Moon. The first part of his letter to Collins is very technical and quotes figures for the orbit. In his last paragraph, he describes the difference between the earlier and the later theories:

> Derby. August 5 1672
> the difference betwixt this first systeme of Mr Horrox in the letter of Dec. 20, 1638, and that collected from his later papers by Mr Crabtree is so wide, that I beleiv you would not thinke that [of 1638] fit to be placed before tables from which they differ so much. I have therefore translated so much from that Epistle of Crabtre as concernes the systeme and trigonometricall method of calculation which you may cause to be printed before the Tables which are framed from them and agree punctually with them[4]

Flamsteed wanted to include his own wording of the lunar theory in the *Opera Posthuma* instead of Horrocks's original. John Wallis was naturally unwilling to substitute a work of Flamsteed in place of one by Horrocks, and as editor he should have been allowed the final word. Flamsteed was insistent, however, and he must have had friends in high places, for when the book eventually appeared in print it was his account which appeared in all except a few of the earliest editions. To be fair to Flamsteed, his main motivation was that he had produced more accurate values for the constants of the motion than Horrocks had achieved forty years earlier.

Another problem arose when the publisher Hickman went broke. He had made a loss on John Wallis's own book and on that of the mathematician Isaac Barrow, who was Newton's tutor at Cambridge. Fortunately, another publisher, Moses Pitt, was prepared to take the work on and he bought a 'remain' of over two hundred copies. At last the work was ready for publication. John Collins wrote to Dr Edward Bernard from his house next to the Three Crowns in Bloomsbury Market:

> Dr. Wallis, his comment on the astronomicall remaines of Horrox, is to goe to the Presse here, and there is now a type provided for the same, the Doctor desired to revise it first, that he might adde a running title to the

> Topp, I sent it on this day three weekes by Dobbins, Moores coachman; giving notice to the Doctor thereof by the Post and since wrote to the Doctor, but receiving no answer am afeard the Doctor is by his disease incapacitated, or under some great affliction.[5]

John Wallis's ailment is a minor mystery, but it cannot have been serious. Other evidence shows that he generally enjoyed good health and lived to be over eighty. John Worthington, however, the friend of both Horrocks and Wallis at Cambridge, unfortunately died the year before *Opera Posthuma* was published. The book at length made its appearance in 1672. The *Philosophical Transactions* published a review, headed by the full, elaborate Latin title:

> *Jeremiae Horroccii Angli Opera Posthuma: una cum Guil, Crabtraei Observationibnus Coelastibus; nec non Joh. Flamstedii de Temporis Aequatione Diatriba, Numerisq; Lunaribus ad novum Lunae Systema Horroccii. Londini, impensis Joh. Martyn, R. Societatis Typographi, A. 1672. in 4.*

The review started with a brief history of the author, describing the publication of *Venus in Sole Visa* and mentioning his extreme youth. It went on to describe the problems he encountered with Lansberg, his discovery of the *Rudolphine Tables* and his success with the system proposed by Kepler. The review includes a useful concise account of the astronomical works of Horrocks:

> This Horrox is the same with him, that is the Author of that excellent Tract, called Venus in Sole visa, publish'd by the famous Johannes Hevelius together with his Mercurius in Sole visus: who if he had not been snatch'd away by an untimely death in the flower of his age, would certainly, by his industry and exactness, which did accompany his great affection to Astronomy, have very considerably advanced that Science. Now we have only left us these imperfect Papers, digested, not without great care and labour, by that Learned Mathematician Dr. John Wallis; Wherein does occurr,
>
> First, the Keplerian Astronomy, asserted and promoted; which this Author undertook, after he had spent much time, and taken great pains in acquainting himself with that Lansbergius, which he at first embraced with to much eagerness and addition, that it was difficult to divorce him

JEREMIÆ HORROCCII,
LIVERPOLIENSIS ANGLI, ex Palatinatu
LANCASTRIÆ,
OPERA POSTHUMA;

viz.
Aſtronomia *Kepleriana*, defenſa & promota.
Excerpta ex Epiſtolis ad *Crabtræum* ſuum.
Obſervationum Cœleſtium Catalogus.
Lunæ Theoria nova.

Accedunt
GUILIELMI CRABTRÆI, *Manceſtrienſis*,
Obſervationes Cœleſtes.

In calce adjiciuntur
JOHANNIS FLAMSTEDII, *Derbienſis*,
De Temporis Æquatione Diatriba.
Numeri ad Lunæ Theoriam *Horroccianam*.

LONDINI,
Typis GULIELMI GODBID, Impenſis J. MARTYN Regalis
Societatis Typographi, ad inſigne Campanæ in Cœmeterio
D. *Pauli*, Anno Domini M. D. C. LXXIII.

FIG. 14. Title page of *Opera Posthuma*.

> from it, till at length, by the advertisements of William Crabtree, a sagacious and diligent Astronomer of that time, he found, that neither the Hypotheses of Lansbergius were consistent among themselves, nor his Tables agreed with Observations exactly made, nor the Precepts of them were well demonstrated or could be, whatever that man boasted of the wonderfull agreement of his Tables with the Observations of former times: All which errors having been found at last by our Author himself, and withall the writings of Kepler, and the Rudolphin Tables by him search'd into, he saw cause far to prefer them to the Lansbergian, because grounded upon Hypothesis consonant to Nature, and well agreeing with the Heavens: though he found cause by his accurate Observations to amend even these Tables, yet without a necessity of changing the Hypothesis. In which work when he was well engaged, he was cut off by death very young, in the 23th year of his age. His first piece then, were his Disputations against the Astronomy of Lansbergius, in which he clearly demonstrates, that the Hypothesis of that Author do neither agree with the Heavens nor among themselves, Which argument he carried on so far, that having finished the four first Disputations (as they are here to be found) he had begun a few sheets of the fifth, which was about the Diagram of Hipparchus, from which some have pretended exactly to demonstrate the Distance of the Sun. After which follow two Disputations more; the one, of the Celestial bodies and their Motion, the other, his Answer to the Cavils Hortensuis against Tycho. So much of the First part of this Volume.[6]

The second part of the publication contained Latin translations of over forty letters written by Jeremiah Horrocks to William Crabtree, his 'intimate friend and industrious companion in the study of astronomy'. The letters included some useful celestial observations, interlaced with valuable discussion concerning the methods used to make the observations. The third part consisted of a catalogue of astronomical observations made by Horrocks before he realized that they should be corrected by allowing for the eccentricity of the eye, but showing the corrections he had calculated. The fourth part consisted of Horrocks's theory of the Moon, 'together with the Lunar Numbers of Mr. Flamstead upon it'.

The astronomical observations of William Crabtree were also included, and in particular, his positions of the planets Saturn, Jupiter, Mars and Venus. This was followed by another contribution from John Flamsteed, a dissertation on the inequality of the solar day due to the eccentricity of the Earth's orbit 'wherein are demonstrated the Prosthaphaereses of the time, necessary to make an Equation, and proceeding from the Unequal motion of the Earth from the aphelion to the perihelion, and the Inclination from the Equinoxes to the Solstices, and vice versa'.

The book was long overdue and it was very well received. Things had moved on since the time of Horrocks, but one advantage was that there were more people around who were able to appreciate what he had achieved in his short working life. In Scotland, it was read by James Gregory of St Andrew's University. He wrote to Collins on 7 March 1673:

> I received these letters ye mention, and also the box together with the things contained, & particularlie Horrocci posthuma, for which I must acknowledge my selfe exceedinglie engaged to you: I have perused him & am satisfied with him beyond measure; it was a great loss that he dyde so young; many naughtie fellows live till 80[7]

It should perhaps be added that Gregory used the word 'naughtie' in its original mode, which meant 'good for nought'. Another member of the Royal Society wrote to John Collins to express his satisfaction:

> Sr
> This day fortnight I received your letter accompanied wth part of the remainder of Mr Horrox ... I am very glad that the world will enjoy the writings of the excellent astronomer Mr Horrox[8]

The writer was a young man from Cambridge University. He had recently succeeded Isaac Barrow as the Lucasian professor of Mathematics. His name was Isaac Newton.

After the publication of *Opera Posthuma*, Flamsteed returned to his home in Derby, where he had time to study a new treatise on the Moon published by a Mr Streete. Flamsteed soon realized that Streete's new treatise was a blatant case of plagiarism. Streete had read the *Opera Posthuma* and had seen that Horrocks's ideas were exactly what he needed to perfect his own theory of the Moon. It is no wonder that the name of Thomas Streete has long been forgotten as an astronomer, but he could very easily have stolen one of Horrocks's prizes. At the time there were very few people alive who could uncover his guilt and it was fortunate that Flamsteed was one such person. He wrote to John Collins about his findings:

> After my return to Derby I fell to peruse Mr Streets discourse, and to consider the contrivance of his moon-wiser, and cannot but blush at the confidence of the man whom I took to have been more ingenuous than to have imposed such things upon us for his, which were never thought of by him, till they were taught him by others plainly; though he affirms in the conclusion of his little tract, that it is different from Mr Horroxes; of whom he seems to speak but very slightly. I can assure that for the motion of Longitude tis the very same, and no other then what he took from my explication, save that, where I thought the manner of Librating the Apogeum was obvious from the Calculation, and needed not be explain'd, he hath shewn, how to take it in the Libratory Circle. For his motion of Latitude, indeed it is a litle different, but I can assure him not much better then Mr Horroxes; to whose memory had he been as Just as I, he could not have boasted, that for this contrivance we were any way indebted to his ingenuity.[9]

Flamsteed had not finished. He went on to say that, even with the plagiarized correction, Streete's method was nowhere near accurate enough to solve the pressing problem of finding the longitude at sea. He calculated that, even after stealing from Horrocks, Thomas Streete could only predict the position of the Moon to half a degree, the diameter of the Moon itself.

Flamsteed was also working on Horrocks's theory of the Moon, but with full acknowledgements. He wanted to simplify the method so that navigators could use it with a minimum of calculation, and he thought he could predict the true position of the moon to within one or two minutes of arc – more accurate than Streete by a factor of thirty. He was also able to improve on the motion of the Sun by using the observations of Cassini at the Paris observatory. 'The Cold weather has hindered me,' he complained, 'Else I had made a good progress towards the completing this contrivance, which I hope to perfect in a short time'[10]

John Flamsteed went to great lengths to defend Horrocks's work, but he knew as well as anybody that when Horrocks came to estimate the parallax of the sun he had used a false assumption. In the seventeenth century, the only way to measure this parallax was by a direct measurement of the distance between two bodies in the solar system. It meant measuring angles to within seconds of arc, and the micrometer was the only practical means by which this could be accomplished. At its closest conjunction to the Earth, the reddish planet Mars was fully illuminated by the Sun and the stars in the background were easy to observe. As the rotation of the Earth carried the observer around during the night, observations could be made from points which were several thousand miles apart and astronomers realized that it might be possible to use this effect to measure a parallax for Mars. John Flamsteed was an active correspondent of Richard Towneley and he made the occasional visit to Towneley Hall to meet his colleague and talk about astronomy. Richard Towneley's telescope was fitted with a Gascoigne micrometer. In 1672 Flamsteed made two visits to Towneley Hall. On the second visit, in September, he obtained the measurements he wanted, using the micrometer. The angles were smaller than he had expected. He wrote to Henry Oldenburg at the Royal Society:

> Last September I was at Towneley the first weeke that I intended to have observed Mars. There with Mr Townley I twice observed him, but could not make two observations as I intended in one night. The first night after that of my returne I had the good hap to measure his distances twice from 2 stars the same night whereby I find that his parallax was very small certeinely not 30 seconds. so that I believe the Suns [parallax] is not more than 10 [seconds].[11]

Mars had its own proper motion as it orbited the Sun, but this was known very accurately. The small discrepancy of about 30 seconds from the proper motion was the parallax he sought. A simple calculation gave him the distance from Earth to Mars and a second calculation gave him the distance from the Earth to the Sun. In February the following year, he had worked over his figures and wrote to tell John Collins at the Royal Society:

> but this I have certeinely learnt from my observations than the Suns parallax is not more than 10": yea probably but 7" and his distance a terra 26000 semidiameters which is a distance to which none ever durst remove him yet and thrice as far as I supposed him formerly in my solar tables[12]

The modern accepted value of the Sun's parallax is 8.783 +/- 0.015 arc seconds. Flamsteed's 26,000 Earth semi-diameters equates to a distance of about 150 million kilometres to the Sun. Flamsteed knew that the world would not believe his estimate and he was right. It was unfortunate for Flamsteed, however, that he did not publish his findings immediately, for in Paris the French astronomers Cassini and Picard measured the parallax of Mars at the same conjunction. They published their findings and were able to claim a priority on the result. Flamsteed was not as bold as Horrocks when it came to shaking off all preconceived notions of the scale of the universe, but he certainly deserves credit for being one of the first astronomers to get close to the correct value for this very difficult and important constant. Credit is also due to Richard Towneley for making the observation possible and adding another scientific milestone to the history of the Towneley family.

CHAPTER TWELVE

Towards Posterity

When the Royal Greenwich Observatory was founded in 1675, John Flamsteed was appointed the first Astronomer Royal. It was a recognition of his talents and observing skills. One of his major tasks was to prepare an accurate stellar catalogue for the stars of the northern hemisphere. The major problem of the times, the motion of the Moon, was very important for commercial reasons. If the problem could be solved, it would provide the key to the problem of finding the longitude at sea. It was well known that if tables could be published to predict the position of the Moon to within 1 minute of arc, navigators would be able to use the tables to calculate their longitude. An accurate theory for the motion of the Moon could therefore provide the solution to the longitude problem and this was the reason why Flamsteed was so interested in the lunar motion. By the time the Royal Greenwich Observatory was created, he was better informed about the problem than any other astronomer and this was one of the main reasons why he was able to obtain the coveted appointment of Astronomer Royal.

In 1684 Edmund Halley made a visit to Cambridge University to ask the Lucasian professor of Mathematics, Isaac Newton, a question regarding the motion of a planet about the Sun under an inverse square law. It was a problem which had been debated at the Royal Society and which nobody was able to solve. To Halley's joy and amazement, Newton answered that the orbit of the planet would be an ellipse and that he had made the mathematical calculation to prove it. Edmund Halley recognized that Newton had solved the key problems of the laws of mechanics and the laws governing the motions of bodies in the universe. He persuaded Newton to write up

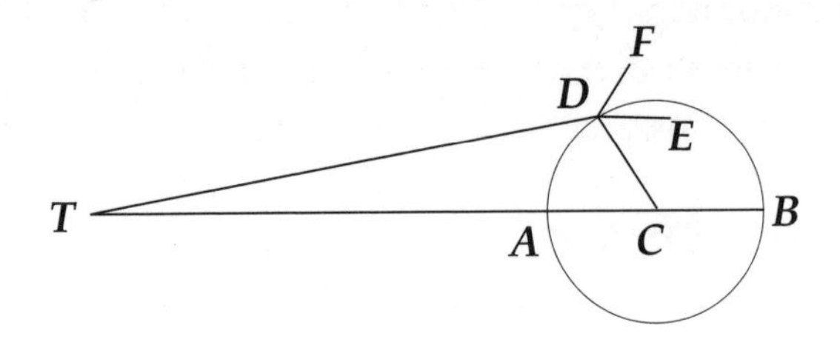

FIG. 15. Newton and Halley's diagram of Horrocks's lunar theory
Newton and Halley used a different diagram (from Fig. 12) to explain Horrocks's theory. The point 'T' is the earth and the point 'D' is the empty focus of the Moon's orbit. The point 'D' moves around a circle, changing the properties of the ellipse in a cyclic fashion.

his work on mechanics and gravitation. The Royal Society agreed to publish Newton's work, but when the members found themselves short of funds it was Edmund Halley who dipped into his own pocket to make sure that Newton's *Philosophiae Naturalis Principia Mathematica* became available to the world.

Newton readily acknowledged the priority of Horrocks's work on the motion of the Moon. He also owed something to Horrocks for his theory of celestial mechanics and the ideas behind gravitation. The extent of Newton's debt in this respect is difficult to measure, however. There were many others who had philosophized about the mechanism responsible for the orbits of the planets. Newton had certainly read the *Opera Posthuma* long before he came to write the *Principia*, but the notion that the force of the Earth on the apple was the same as the force which held the Moon in orbit did not come from *Opera Posthuma*. The idea came to him in 1665, the year of the great plague, in his garden at Woolsthorpe. This was several years before Horrocks's thoughts on celestial mechanics were published.

Once the universal principle of the theory of gravitation had been established, it should have been possible to create a perfect system to predict the motion of the Moon based on the gravitational attractions of the Earth and the Sun. It was not so easy. Newton worked on the problem for many years, long after the publication of the *Principia*. To perfect his theory, he became dependent on John Flamsteed to provide astronomical data on the positions of the Moon. Newton and Flamsteed had studied all the available theories on the lunar motion in their attempts to solve the problem. Newton agreed with Flamsteed that Horrocks's theory of the Moon was closest to the truth, but Flamsteed's earlier claims for the theory were too optimistic and in the worst cases it could still be in error by as much as 8 minutes of arc. Newton continued to work on the problem, but it did not fall to his mathematics and he admitted that the Moon was the only problem which made his head ache. His system was to add further 'equations' or corrections to the motion. The Horrockian theory of the Moon, as it became known, was still in use well into the eighteenth century. In the year 1723, Thomas Hearn wrote a passage in his pocket notebook:

> Mr Horrox, a young man, minister of Hoole, a very poor pittance, within four miles of Preston, in Lancashire, was a prodigy for his skill in astronomy, and had he lived, in all probability, he would have proved the greatest man in the whole world in his profession. He had a very strange unaccountable genius, and he is mentioned with great honour by Hevelius upon account of his discovery of Venus in the Sun, upon a Sunday; but being called away to his devotions, and duty at church, he could not make such observations, as otherwise he would have done. He died very young, I think at about 23 or 24 years of age. His posthumous works were printed by Dr. Wallis. They are now scarce. Mr. Whiteside of the Museum, bought them several years agoe, but gave 7s 6d for them[1]

The passage has been copied and repeated many times and it represents the common view of Jeremiah Horrocks and his observation of the transit of Venus. It was written long after the death of Horrocks and it is wrong in many respects, but it seems to be the source of most of the misapprehensions about him. He was not the minister of Hoole. He was not even the curate. The living at Hoole was not a particularly poor pittance. The distance to Preston is seven or eight miles. The spelling 'Horrox' is the Latinized version of the name, allowed by Wallis in the *Opera Posthuma*, but strictly speaking the name should be 'Horrocks' for a piece written in English. The account is correct in every other respect. Jeremiah Horrocks did observe Venus on the Sun's disc and he did make the observations at Hoole. They took place on a Sunday. He did die very young, even younger than the age quoted by Hearne. He certainly did have a strange, unaccountable genius.

Biographies were rare in the eighteenth century, particularly when the subject was a scientist or an astronomer. The *Opera Posthuma* and the *Venus in Sole Visa* were, however, available to astronomers, and they were able to keep the knowledge of Horrocks's achievements alive within their own circle. It was not until the 1760s that the next transits of Venus took place and astronomers could once again observe the sight first seen by Horrocks and Crabtree. The second in this pair of transits was observed from the Pacific island of Tahiti in 1769 by Charles Green and Captain Cook on the voyage of the *Endeavour*.[2] This was the long-awaited opportunity to measure the solar parallax using a method devised by Edmund Halley. The attempts to measure the exact contact times of Venus on the Sun's disk were frustrated by the atmosphere of Venus. This produced a 'teardrop' effect which made it impossible to be sure of the precise start and end time of the transit. It will be remembered that during the 1639 transit, when Jeremiah Horrocks was called away on 'business of the highest importance', he did not observe the start of the transit and he did not see the teardrop effect. Had he done so, he might have saved the eighteenth century astronomical community a great deal of frustration.

The nineteenth century saw a revival of interest in Jeremiah Horrocks and in 1859 a biography was published by the Revd Arundell Blount Whatton. He was able to draw on material collected by his father, W. R. Whatton, who had himself been collecting information for a biography but died before he could publish it. Whatton's biographical material covered about a hundred pages, but he also published a valuable English translation of Horrock's own work, *Venus in Sole Visa*. This was the treatise published by Hevelius in 1662. It included Jeremiah's poetry and many of his other ideas about the solar system.

At a later date, memorials to Jeremiah Horrocks were established at St Michael's Church in Dingle and at Richard Mather's ancient chapel. At the church of Hoole, a chapel was built to his memory with a stained-glass memorial window. The sundial, on which Horrocks is reputed to have carved the words '*Ut hora, sic vita*' and '*Sine sole sileo*', was replaced in the nineteenth century with a clock bearing the same inscription. There were another pair of transits of Venus in the 1880s and these raised a great deal of interest amongst the Victorians. The country could hardly have responded better to the occasion and Jeremiah Horrocks was given a memorial in Westminster Abbey. He is described as the curate of Hoole, but otherwise the wording describes his achievements accurately.

In Memory of
JEREMIAH HORROCKS
Curate of Hoole in Lancashire,
Who died on the 3d of January, 1641, in or near his
22d year;
Having in so short a life
Detected the long inequality in the mean motion of
Jupiter and Saturn;
Discovered the orbit of the moon to be an ellipse;
Determined the motion of the lunar apse,
Suggested the physical cause of its revolution;
And predicted from his own observations the
Transit of Venus
Which was seen by himself and his friend
William Crabtree
On Sunday, the 24th of November (O. S.) 1639:
This tablet, facing the Monument of Newton,
Was raised after the lapse of more than two centuries
December 9, 1874

The tablet does indeed face the memorial to Newton, but it is fifty metres away, on the right of the west door and near the burial place of the Unknown Warrior.

The twentieth century produced relatively few publications on the works of Horrocks. Excellent work has been done by Sydney Gaythorpe and Alan Chapman, but these amount to articles in the astronomical press rather than popular biographies. An M.Sc. thesis by Betty Davis of London University appeared in 1967, drawing mainly on the *Opera Posthuma*, and it is still seen as one of the best accounts of his work. In the twentieth century there was no transit of Venus with which to raise interest, but when the craters on the Moon were given their names, Horrocks was remembered again. Between the Sea of Tranquillity and the Ocean of Storms lies Crater Horrocks, thirty kilometres in diameter, just below the Moon's equator and near the centre of the full Moon.

The question of where Jeremiah Horrocks stands in the history of his chosen subject needs to be addressed. The traditional view is to see him as a brilliant youth, far ahead of his time, who was cut off in his prime before he could reach his full potential, and this is a difficult viewpoint to challenge when all the evidence is taken into account. He never grew old, he was always young and full of enthusiasm. He did not live long enough to reach the age where he became disillusioned, cantankerous and discontented about life. There is a school of thought, however, which claims that Horrocks was not entirely unfortunate and in some respects he was actually very lucky. An astronomical event that occurs less frequently than once a century happened at exactly the right time for him. But then we need to ask the question, does genius bring its own luck? It seems obvious that had he lived longer he would have made more major discoveries, but we also know that even some of what he did manage to achieve in his short life has been lost. How much was lost? It does not seem possible that a man so young could have achieved more than he did in the time available to him. His project at the time of his death was to try to formulate a theory of the tides, and we know from his other work that he was perfectly capable of producing something valuable from his observations. Had he lived longer, he might well have tackled the problem of the marine chronometer to keep good time at sea. He already knew more than most about mechanical timekeepers. But what he might have achieved can never be more than speculation. Only what he actually did achieve is pertinent.

Whatton quotes a passage by a Dr Tatham, who described Horrocks as the forerunner of Newton. He brings out the point that Horrocks saw that the way forward was by experiment and observation. This was something which greatly impressed the Royal Society.

> That every philosopher has an absolute right to avail himself of the labours and discoveries of his predecessors, as a legacy freely given him, is a privilege which philosophy itself always claims. It is however a tribute justly due to the memory of this extraordinary genius, Mr. Horrox, whilst we regret the loss of many of his valuable works, to acknowledge from what has been saved, that he was principally instrumental in calling philosophy out of the regions of fictitious invention, and putting her on the investigation of the physical causes of things from experiments and observations; that he not only made the applications of projectile motion to the analogical illustration of celestial, but also assigned the forces of projective and attractive, on which all geometrical calculations are founded; and that, without injuring the immortal fame of his great successor, he may be fairly considered the forerunner of Newton.[3]

Tycho Brahe was the greatest of the naked-eye observers, but he contributed very little to astronomical theory. Kepler, the great theorist, was a poor observer. His observation of the transit of Mercury was a complete disaster. Newton, the greatest theoretician, had to rely on Flamsteed for accurate observations of the moon and the planets. Galileo was a great observer. He did much for terrestrial mechanics, but little to advance the theory of celestial mechanics. Horrocks was the complete astronomer. For a few short years, he was both the best observer and the best theoretician of his times.

It is impossible to write about Horrocks without making comparisons with Newton. Isaac Newton wrote millions of words. Jeremiah Horrocks wrote a few thousand. Newton was an early developer. His *annus mirabilis* was the year 1665, when he was twenty-two years old, but the earliest of his papers dates from later in the 1660s and he was forty-five when the *Principia* was published. Had Newton died at twenty-two, as did Horrocks, he would be almost unknown. It proves nothing, but it is a sobering thought. There are differences in personality, but some of these are due to the fact that where we always see the enthusiastic youngster in Horrocks, we see the mature and measured man in Newton's writings. There was one obvious personality difference. Whenever he was moved by his discoveries and by the works of nature, Horrocks would express his feelings by writing an ode in praise of his mentors or his discoveries. Newton had no time for such frivolities. Only the Persian philosopher Omar Khayyam is the equal of Jeremiah Horrocks as an astronomer poet.

Perhaps the best way to try to assess his life and works is through the opinions of his fellow astronomers, those who came to study the sky after him. His contemporaries had no doubt about his genius. Samuel Foster of Gresham College spoke of him as 'a genius of the first rank'. Flamsteed, too, as he searched high and low to try to find Horrocks's undiscovered manuscripts, had no doubts about his merits. Newton was more reserved when he described Horrocks as 'an excellent astronomer'. Of those who followed, we must mention William Herschel, who was voicing the opinion of his times when he claimed that Jeremiah Horrocks was 'the pride and boast of British astronomy'.

There is one more thing to bear in mind. Simply that Jeremiah Horrocks was the first in his own country. His work may seem primitive alongside that of Newton and Halley, but the later astronomers did not see it as such. They appreciated that it was executed at a time when instrumentation was much more basic, and they looked to him for inspiration. The reaction to the publication of Wallis's *Opera Posthuma* in 1672, and the fact that the Royal Society went to so much trouble and expense to publish it, shows the high esteem in which he was held. Horrocks was preceded by Kepler. Kepler was preceded by Tycho Brahe. Tycho was preceded by Copernicus. During these times, the British had made no significant contribution to astronomy. As regards English astronomy, a handful of stargazers and astrologers preceded Horrocks but there are no astronomers of world ranking. Jeremiah Horrocks was the first in his own country. His work was executed more than a generation before the foundation of the Royal Society. His observations were made nearly forty years before the creation of the Royal Greenwich Observatory. His ideas were formulated two generations before the publication of Newton's *Principia*. Jeremiah Horrocks hardly came of age, yet he is remembered as the father of English astronomy. 'If I have seen further than others before me,' said Newton, 'it is because I have stood on the shoulders of giants.'

Jeremiah Horrocks was one of those giants.

HORROCKS OF BOLTON AND TOXTETH FAMILY TREE

Names in **bold** appear in the text

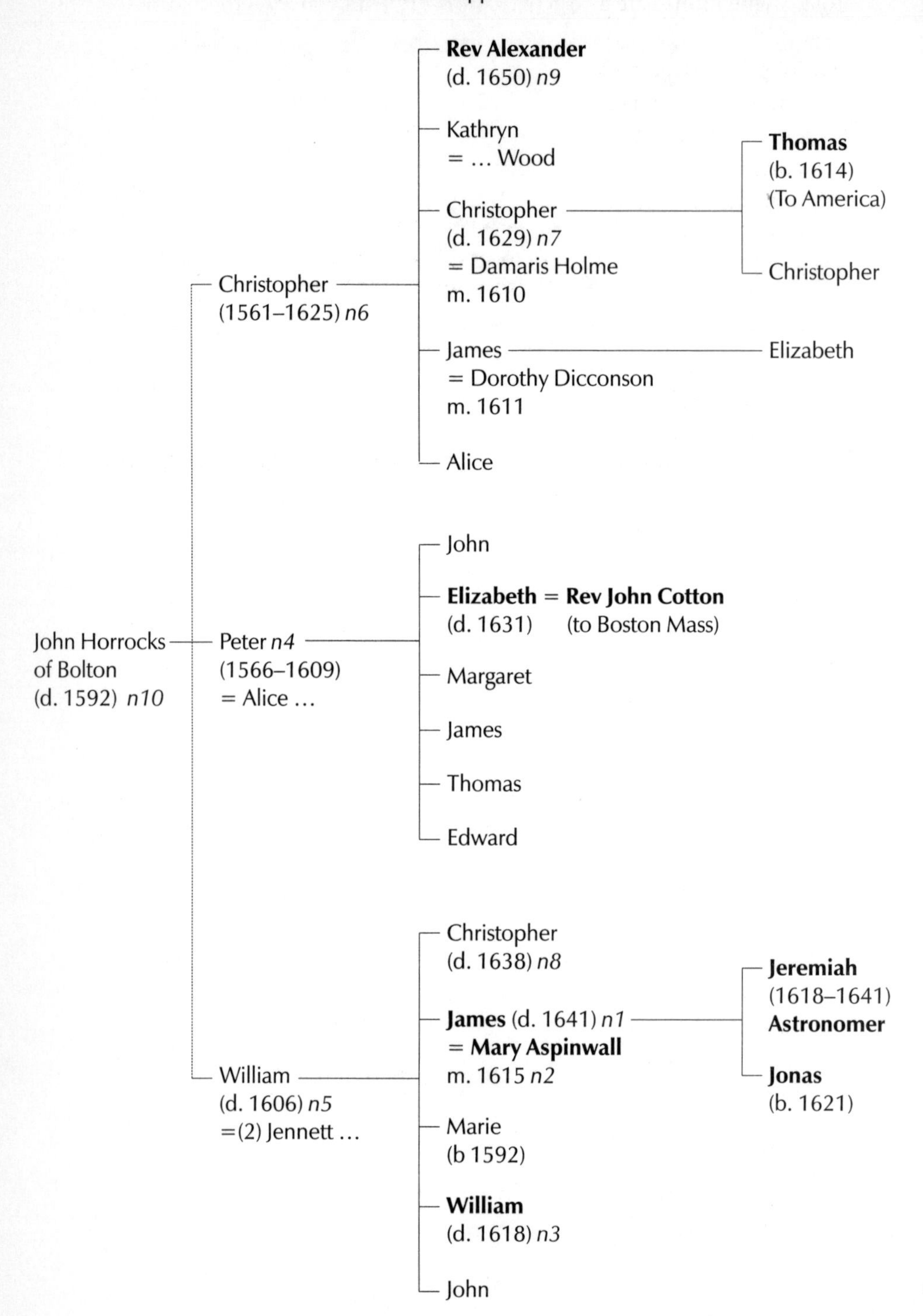

All wills mentioned are held at the Lancashire County Records Office.

n1: John Flamsteed recorded that Jeremiah's father died on 3 May 1641, 'griefe for his sonnes hastening his own death'. The bishop's transcripts for St Nicholas Chapel, Liverpool record the burial of 'James Harrocks, watchmaker' in May 1641. We are therefore indebted to John Flamsteed for the information that James Horrocks, the watchmaker who married in 1615, was the father of Jeremiah.

n2: parish register of Deane. Marriage of James Horrocks and Mary Aspinwall, 16 January 1615.

n3: will of William Horrocks of Toxteth Park, d. 1618. Wife Jennet (also called Joane), brother John Horrocks, cousin Alexander Horrocks (preacher), John my son (under twenty-one). Possibly the schoolmaster to Richard Mather.

n4: will of Peter Horrocks of Edge, d. 1609. Wife Alice. Children: John my eldest son, Elizabeth (under twenty-one), Margaret (under twenty-one), James, Thomas, Edward.

n5: will of William Horrocks of Rumworth (Dean), d. 1606. It seems likely, but is unproved, that William Horrocks was Jeremiah's paternal grandfather. The evidence rests on the fact that he had a son called James who was the right age to be Jeremiah's father. Wife Jennet. Children: James, Jane wife of George Pedleburie, William (under eight), Marie (under fourteen). Possible grandfather of Jeremiah Horrocks.

n6: will of Christopher Horrocks the elder of Sharples, d. 1625. Executor Christopher Horrocks the younger. Debt owed by Thomas Aspinwall.

n7: will of Christopher Horrocks of Turton, d. 1629. Brothers: Alexander and James. Children: Christopher, Thomas.

n8: will of Christopher Horrocks of the Foulds, d. 1638. James Horrocks of Toxteth Park, watchmaker, my brother. Alexander Horrocks my cousin. James, Christopher and Alice children of Christopher Horrocks my late uncle deceased.

n9: will of Revd Alexander Horrocks, d. 1650. Kathryn Wood my sister, Elizabeth Holme my sister, Elizabeth Horrocks my careful niece. Thomas Horrocks my nephew, the son of Christopher Horrocks. Elizabeth Horrocks daughter of my late brother James Horrocks.

n10: John Horrocks of Bolton le Moors, d. 13 May 1592. His son was Christopher Horrocks, aged thirty-one. Victoria County History, Lancashire, vol. 7, p. 279 n. 45

Alumni of Cambridge:

Horrocks, John: Emmanuel, 1589
Horrocks, Thomas: Magdalene, 1593
Horrocks, Thomas: St John's, *c.* 1593
Horrocks, John: Christ's, 1612
Horrocks, William: Clare, 1614
Horrocks, John: Emmanuel, 1622
Horrocks, Thomas: St John's, 1631, aged seventeen
Horrocks, Jeremiah: Emmanuel, 1632

References:

Jeremiah Horrocks: date of birth, parentage and family associations, Transactions of the Historical Society of Lancashire and Cheshire, 106 (1954), 23–33. A very valuable contribution, but there are several errors in Gaythorpe which have been repeated by other researchers. Most of these have been generated by the confusion of Thomas Aspinwall (1571–1612) with his son Thomas Aspinwall (d. 1624). The latter clearly states in his will that he was cousin (not uncle) to Jerehiah Aspinwall.

ASPINWALL OF TOXTETH FAMILY TREE

Names in **bold** appear in the text

- James (d. 1591) Son of William Aspinall = Katherine... m. 1538 (d. 1596) *n4*
  - William = Katherine Barker m. 1564 *n4*
    - Peter (b. 1568) = Elizabeth Fletcher m. 1596
    - **Edward** = Anne ...*n4&6* 1567–1633
      - Jerehiah (b. 1595) = Mary Cowper m. 1615
        - **Edward** (d. 1656) *n5* = Eleanor Ireland (b. 1622)
      - Peter
    - Elizabeth (b. 1570)
    - James 1570–1632
    - **Thomas** 1571–1612 *n4* = Mary... m. 1589
      - Elizabeth = Peter Ambrose
      - **Thomas** (d. 1624) *n1* = Margaret...
        - **Samuel** (d. 1672) *n3* = Jane...
        - **Elizabeth** = **Christopher Horrocks** *n2*
        - **Peter** (to America) (b. 1612) *n7* = Alice Sharp m.1645
      - **Mary** = **James Horrocks** [Parents of **Jeremiah**] m. 1615
  - Anne
  - Grace = Thomas Greaves m. 1591
  - Edward = Elizabeth Parker
  - James (d. 1624) = Elizabeth Fairclough m. 1570 *n4*
    - Edward (b. 1571)
    - Thomas (b. 1573)

All wills mentioned are held at the Lancashire County Records Office.

n1: will of Thomas Aspinwall of Toxteth Park, d. 1624. My brother in law James Horrocks. My cousin Jerehiah Aspinwall. My sister Elizabeth Ambrose. Elizabeth wife of Peter Aspinwall. My wife Margaret.
n2: will of Christopher Horrocks of Warrington, watchmaker, d. 1663. My wife Elizabeth, my brother Samuel Aspinwall.
n3: will of Samuel Aspinwall, d. 1672. My sons Isaac, Josia, Timothy. My wife Jane. My brother Peter Aspinwall. My sister Elisabeth Horrocks.
n4: Ormskirk Parish Registers. Marriages: James Aspinall to Katherine, 1538; William Aspinall to Katherine Barker, 1564; James Aspinwall to Elizabeth Fairclough, 1570. Baptisms: Edward, son of William and Katherine Aspinwall, baptized 1567; Thomas, son of William and Katherine Aspinwall, baptized 24 March 1571.
n5: Dugdale's Visitation 1664–5, Chetham Society, vols 84, 85, 88.
n6: Duchy of Lancaster Inq p m, vol. 28, no. 1. Edward Aspinall of Toxteth Park.
n7: Married at Dedham, Mass., *c.* 1645. Descendants in America.

Alumni of Cambridge:

Aspinall, James: St John's, 1564.
Aspinall, Gamaliel: Clare, sizar, 1582.
Aspinall, Thomas: Gonville and Caius, 1589, aged fifteen (of Norwich).
Aspinwall, Edward: Emmanuel, 1634, aged fourteen.

Alumni of Oxford:

Aspinall, Thomas: (no college given), 1542.
Aspinwall, Alexander: Brasenose, 1546.
Aspinall, Brian: 1603, (pleb), aged fifteen.
Aspinall, Miles: St Mary Hall, *c.* 1581, aged twenty-three.
Aspinall, Edward: Brasenose, 1585, aged fifteen.
Aspinall, Alexander: Brasenose, 1601.
Aspinall, William: Brasenose, 1621, aged eighteen.
Aspinall, Timothy: Brasenose, 1620.
Aspinall, Peter (son of Edward): Brasenose, 1636.
Aspinall, Thomas: Brasenose, 1650.

Aspinall, Peter : Brasenose, 1655, aged fifteen.
Aspinall, Gilbert (son of Edward): Christ Church, b. 1647.
Aspinall, Edward (son of Edward): Christ Church, b. 1648.

References:

Aspinall, Henry Oswald: *The Aspinwall and Aspinall families of Lancashire* 1189–1923 (Exeter, 1923). This book is another very valuable contribution, but it is incomplete regarding the Jeremiah Horrocks connections. Henry Aspinall's book does not mention Jonas Moore, the man remembered for draining the fens and for his contribution towards the founding of the Royal Greenwich Observatory. Jonas Moore may well have been related to Jeremiah Horrocks and he was almost the same age. Both their mothers were called Mary Aspinwall.

CHRONOLOGY

Year	*Horrocks*	*Astronomy*	*Other Events*
3rd century BC		Aristarchus of Samos	
2nd century BC		Hipparchus	
2nd century AD		Ptolemy	
1543		Copernicus's *De Revolutionibus* published	
1591	Toxteth was 'de-parked'		
1601		Tycho Brahe's tables published	
1612	Richard Mather's school, Toxteth		
1615	James Horrocks married Mary Aspinwall		
1618	Richard Mather's Chapel was built		
1618 Jan–May	Birth of Jeremiah Horrocks		
1619		Kepler's *Harmonice Mundi*	
1620			Voyage of the *Mayflower*
1621	Birth of Jonas Horrocks		
1626			Death of Francis Bacon
1630			Winthrop's fleet sailed to Massachusetts
1632	Matriculated Emmanuel College		
1632		Galileo's *Dialogue*	
1632		Kepler's *Rudolphine Tables* published	
1635	Returns to Toxteth		
1636	First letter to William Crabtree		
1637	Anti-Lansberg Treatise		
1638	First lunar theory developed		
1638 Dec	Observed lunar eclipse		
1639 May	Observed solar eclipse		
1639 June	Left Toxteth for Hoole		

1639			Foundation of Harvard College
1639 Nov	Transit of Venus		
1640 April	Returns to Toxteth		
1640	Final form of the lunar theory		
1640	Solar parallax		
1640	Crabtree writes to Gascoigne		
1640	Horrocks starts to study the tides		
1641 Jan	Death of Jeremiah Horrocks		
1641 May	Death of James Horrocks		
1642			Outbreak of the English Civil War
1642		Death of Galileo	
1642 Dec		Birth of Isaac Newton	
1644 July			Battle of Marston Moor
1644	Death of Crabtree and Gascoigne		
1649			Execution of Charles I
1653	Shakerley publishes extracts from Horrocks's work		
1660			Restoration of the monarchy
1662	*Venus in Sole Visa* published		
1666			Great Fire of London
1672	*Opera Posthuma* published		
1672	Flamsteed visits Towneley Hall		
1675		Royal Greenwich Observatory founded	
1686		Newton's *Principia* published	
1761 and 1769		Transits of Venus	
1859	Whatton's biography of Horrocks published		
1874 and 1882		Transits of Venus	
1884	Memorial in Westminster Abbey		
2004 and 2012		Transits of Venus	

GLOSSARY

Anomaly: the angular distance of a planet or satellite from its last perihelion or perigee, so called because the first irregularities of planetary motion were discovered in the discrepancy between the actual and the computed distance.

Antecedence: a motion from a later to an earlier sign of the zodiac, or from east to west. Retrograde motion; also a position more to the west.

Aphelion: the point of a planet or comet's orbit at which it is farthest from the Sun. Opposite to perihelion.

Apogee, or *apogeon*: the point in the orbit of the Moon, or of any planet, at which it is at its greatest distance from the Earth. Also, the greatest distance of the Sun from the Earth.

Azimuth: an arc of the heavens extending from the zenith to the horizon, which cuts it at right angles. The quadrant of a great circle of the sphere, passing through the zenith and nadir.

Conjunction: the lining up of three or more bodies. For example, Earth–Venus–Sun, which gives rise to the transit of Venus across the sun.

Consequence: a motion from an earlier to a later sign of the zodiac, or from west to east.

Declination: the angular distance of a heavenly body (north or south) from the celestial equator, measured on a meridian passing through the body. It corresponds to latitude on the Earth.

Ecliptic: the plane of the Earth's orbit extended to infinity from the Sun. So called because eclipses can happen only when the moon is on or very near this plane.

Emersion: the reappearance of the Sun (or the Moon from shadow) after an eclipse, or of a star or planet after occultation. The opposite of immersion.

Equation: the action of adding to or subtracting from any result of observation or calculation such a quantity as will compensate for a known cause of irregularity or error.

Equinoctial: the celestial equator, i.e. the projection of the Earth's equator on to the sphere of the stars. So called because, when the Sun is on it, nights and days are of equal length in all parts of the world.

Equinoctial points or *equinoxes*: the two points at which the Sun's path crosses the Equator (described technically as the first points in Aries and Libra).

Immersion: the disappearance of a celestial body behind another or into its shadow, as in an occultation or eclipse. The opposite of emersion.

Latitude, celestial: the angular distance of a heavenly body from the ecliptic. Thus, the latitude of a star is the arch of a great circle made by the Poles of the ecliptic.

Libration of the moon: an apparent irregularity of the moon's motion, which makes it appear to oscillate in such a manner that the parts near the edge of the disk are alternately visible and invisible.

Libratory: having an oscillating motion like that of the beam of a balance.

Longitude, celestial: the distance in degrees reckoned eastward on the ecliptic from the vernal equinoctial point to a circle at right angles to the ecliptic through the heavenly body, or the point on the celestial sphere whose longitude is required.

Meridian: if a circle were drawn through any place on the Earth's surface and the North and South Pole, it would be called the meridian of that place.

Nadir: a point in the heavens diametrically opposite to some other point. For example, the point of the heavens diametrically opposite to the zenith.

Node: one of the two points at which the plane of an orbit intersects the ecliptic, or in which two great circles of the celestial sphere intersect each other.

Occultation: hiding, concealment from view by something interposed. Hence the concealment of one heavenly body by another passing between it and the observer.

Opposition: the opposite of conjunction. Mars is in opposition at its closest approach when it lines up with the Earth and the Sun.

Orb: a circle, or anything of circular form, such as an orbit or a sphere. The hollow concentric spheres thought in ancient times to carry the planets and stars.

Parallax: the apparent displacement, or difference in the apparent position, of an object, caused by actual change (or difference) of the position of the point of observation. In astronomy, there are two kinds of parallax, diurnal and annual, the former when a celestial object is observed from opposite points on the Earth's surface, the latter when observed from opposite points of the Earth's orbit. The horizontal parallax is the diurnal parallax of a heavenly body seen on the horizon.

Perigee, or *perigeon*: the point in the orbit of a planet at which it is nearest to the Earth. In the Ptolemaic astronomy, applicable to any planet, but now usually restricted to the moon.

Perihelion, or *perihelium*: that point in the orbit of a planet or other heavenly body at which it is nearest to the Sun. The opposite to aphelion.

Precession (of the equinoxes): the earlier occurrence of the equinoxes in each successive sidereal year, due to the retrograde motion of the equinoctial points along the ecliptic, produced by the slow change of direction in space of the Earth's axis.

Prosthaphaeresis: the equation of the centre. The difference between the true and mean place of a planet, or between the true and equated anomaly. The correction necessary to find the true position from the mean.

Quadrature: one of the two points (in space or time) at which the Moon is 90 degrees distant from the Sun, midway between the points of conjunction and opposition. Also used to describe the position of one heavenly body (e.g. the Moon) relative to another (e.g. the Sun) when they are 90 degrees apart from the observation point.

Quartile: the aspect of two heavenly bodies which are 90 degrees distant from each other.

Radius vector: a variable line drawn to a curve from a fixed point as origin. In astronomy, the origin is usually at the Sun or a planet around which a satellite revolves.

Right ascension (of the Sun or a star): the degree of the equinoctial or celestial Equator, reckoned from the first point of Aries.

Sesquialter: a ratio of one-and-a-half to one.

Solstice: one of the two times in the year, midway between the two equinoxes, when the Sun, having reached the tropical points, is farthest from the Equator and appears to stand still, i.e. about 21 June (the summer solstice) and 22 December (the winter solstice).

Superficies: a magnitude of only two dimensions, having only length and breadth. The outer surface of a body which is apparent to the eye.

Syzygy, or *syzigies*: the conjunction or opposition of the heavenly bodies, or either of the points at which these take place. In the case of the Moon with the Sun, the full Moon or opposition is that state in which the whole disc of the moon is enlightened. The new Moon or conjunction is that state in which the whole surface or disk turned toward us is dark.

Zenith: the point of the sky directly overhead. The highest point of the celestial sphere viewed from any particular place. The upper pole of the horizon (opposite the nadir). Loosely, the expanse of the sky overhead, the upper region of the sky, the highest or culminating point of a heavenly body.

NOTES

Abbreviations used:

CS: Chetham Society
LRO: Lancashire Records Office
PRO: Public Records Office
RGO: Royal Greenwich Observatory
THSL&C: Transactions of the Historic Society of Lancashire and Cheshire
VCH: Victoria County History

CHAPTER 1: The Royal Society

1 *Opera Posthuma*, Preface, p. ij.
2 CS, vol. XIII (1847), p. 366.
3 Whatton, p. 63.
4 Ibid., p. 64.
5 Ibid., pp. 64–5.
6 Ibid., pp. 68–9, taken from Bailey's biography of Flamsteed.
7 Rigaud, vol. 2, pp. 118–19.

CHAPTER 2: Toxteth

1 VCH, vol. 2, p. 62.
2 Chandler, p. 382.
3 Hall II, p. 250.
4 Camden, *Brittania*.
5 LRO, DDM 50/3.
6 Britten, p.174.
7 CS, vols lxxxiv, lxxxv, lxxxviii.
8 Twemlow, J. A. (ed.).
9 Halley.
10 Hall I, p. 37.

CHAPTER 3: Cambridge

1 Fuller, p.147.
2 Bendall, Brooke and Collinson, p. 30.
3 Ibid., p. 59.
4 Whatton, p.13 (translated from *Opera Posthuma*).
5 Bush and Rasmussen, chapter 3.
6 Ibid.
7 Scriba, p. 27.
8 Bacon, Book 1.
9 Whatton, p. 9.

CHAPTER 4: Astronomy before Horrocks

1 Copernicus, p. 7.
2 Ibid., p. 8.
3 Ibid., p. 7.
4 Brahe.
5 Ibid.
6 Kepler.

CHAPTER 5: Horrocks Studies the Heavens

1 Bailey, *Transit*, pp. 256–7.
2 Horrocks to Crabtree, 23 November 1637, *Opera Posthuma*, p. 298.
3 Horrocks to Crabtree, 21 June 1636, ibid., p. 247.
4 Whatton, p. 164.
5 Horrocks, *Venus in Sole Visa*, pp.

164–7.
6 Horrocks, 'Philosophical Exercises', S 21.
7 Ibid., S 22.
8 Ibid., S 23.
9 Ibid., S 13.
10 Horrocks, 'Astronomical Exercises', S 8.
11 Horrocks, *Venus in Sole Visa*, p. 109.

CHAPTER 6: Gravitation and Mechanics

1 *Opera Posthuma*, p. 307.
2 Horrocks, *Venus in Sole Visa*, pp. 119–21.
3 Horrocks, 'Astronomical Exercises', S 29.
4 Horrocks, 'Philosophical Exercises', S 9.
5 Ibid., S 10.
6 Ibid., S 16.
7 Horrocks to Crabtree, 25 July 1638, *Opera Posthuma*, p. 310.
8 Whatton, p. 101.
9 Ibid., p. 27.
10 Ibid., pp. 27–9.
11 *Opera Posthuma*, Disp. VI. Cap. 1.
12 Newton, *Philosophiae Naturalis*, Def. V.
13 *Opera Posthuma*, p. 311.

CHAPTER 7: Transit of Venus

1 Whatton, p. 7.
2 Gaythorpe IV, p. 25.
3 Chapman II, p. 349.
4 Horrocks, 'Philosophical Exercises', p. 26.
5 Ibid., Part 1, S7.
6 CS, vol. VII, pp. 4–5 (1841).
7 VCH, *Lancashire*, vol. VI, p. 103.
8 Horrocks, *Venus in Sole Visa*, pp. 127–8.
9 Horrocks to Crabtree, 26 October 1639, *Opera Posthuma*, p. 331.
10 Horrocks *Venus in Sole Visa*, p. 117.
11 Ibid., p. 135.
12 Gaythorpe III, p. 312. Gaythorpe VI, p. 61.
13 Horrocks, *Venus in Sole Visa*, p. 129.
14 Ibid., p. 131.
15 Ibid., p. 114.

CHAPTER 8: The Parallax of the Sun

1 Horrocks, 'Astronomical Exercises', S10.
2 Ibid., S14 *et seq.*
3 Ibid., S15.
4 Ibid., S16.
5 Horrocks, *Venus in Sole Visa*, p. 210.
6 Horrocks, 'Philosophical Exercises', Part 2, S5.
7 Horrocks, *Venus in Sole Visa*, p. 211.
8 *Opera Posthuma*, p. 164.
9 Horrocks, *Venus in Sole Visa*, p. 212.
10 Ibid., p. 201.
11 Ibid., pp. 214-16.
12 Horrocks, 'Astronomical Exercises', S6.
13 Horrocks, 'Philosophical Exercises', Part 2, S5.

CHAPTER 9: The Motion of the Moon

1 Horrocks, *Venus in Sole Visa*, p. 177.
2 Ibid., pp. 177–9.
3 Horrocks, 'Philosophical Exer-

cises', S2.
4 Ibid., S5.
5 Ibid., S14.
6 Newton, *Philosophiae Naturalis*, III, scholium to proposition 35.
7 *Opera Posthuma*, p. 309, Letter of 25 July (see Davis, p. 48).
8 Newton, *Philosophiae Naturalis*, III, scholium to proposition 35.
9 Newton, *Correspondence*, IV, p. 3 (translated from the Latin).
10 Crabtree to Gascoigne, Whatton, pp. 53–4.
11 Crabtree to Gascoigne, ibid., pp. 54–5.
12 *Opera Posthuma*, p. 53.
13 Crabtree to Gascoigne, Whatton, p. 55.
14 Horrocks, 'Philosophical Exercises', Part 2, S 9.
15 Ibid., S 7.
16 Horrocks to Crabtree, 12 December 1640, *Opera Posthuma* p. 337.
17 Horrocks to Crabtree, 19 December 1640, ibid., p. 339.

CHAPTER 10: War

1 *Opera Posthuma*, p. 338.
2 Ibid., p. 335.
3 Gaythorpe IV.
4 Webster, p. 62, note 17.
5 Adair, p. 142.
6 Whatton, p. 107.

CHAPTER 11: *Opera Posthuma*

1 Chapman, p. 3 .
2 Flamsteed, vol. 1, p.180.
3 Ibid., p. 58.
4 Ibid., p. 176.
5 Whatton, p. 70.
6 *Philosophical Transactions*, 1673, pp. 5,078–9.
7 Newton, *Correspondence*, I, p. 259.
8 Ibid., p. 161.
9 Flamsteed to Collins, 25 November 1674, Flamsteed, vol. 1, p. 313.
10 Ibid.
11 Flamsteed to Oldenburg, 16 November 1672, 'Pax de Mars', ibid., p. 185.
12 Flamsteed to Collins, 20 February 1673, ibid., p. 194.

CHAPTER 12: Towards Posterity

1 Hearne's MS Diary, Saturday 8 February, vol. 102, p. 62, (Bodleian Library, Oxford).
2 Aughton, chapters 4 and 5.
3 Whatton, p. 90.

BIBLIOGRAPHY

Adair, John, *By the Sword Divided* (Stroud, 1998).

Aspinall, Henry O., *The Aspinwall and Aspinall families of Lancashire, 1189–1923* (Exeter, 1923).

Aughton, Peter: *Endeavour* (London, 2002).

Bacon, Francis, *Advancement of Learning, Book 1* (London, 1605).

Bailey, John E., *Jeremiah Horrox and William Crabtree, observers of the Transit of Venus, 24 Nov 1639 . . . Palatine Notebook of Dec. 1882*, pp. 253–66 (Manchester, 1883) (Bailey, *Transit*).

——, *The Writings of Jeremiah Horrox and William Crabtree. . . Reprinted, with Additions, etc. from the Palatine Notebook of Dec. 1882, and Jan. 1883* (Manchester, 1883) (Bailey, *Writings*).

Baily, Francis, *An Account of the Revd John Flamsteed, the First Astronomer-Royal* (London, 1835), p. lxxiii.

——, *Supplement to the Account of the Revd John Flamsteed* (London, 1837), pp. 680–93.

Bailey, F. A. and Barker, T. C., *The Seventeenth-Century Origins of Watch-making in South-west Lancashire*, in J. R. Harris, *Liverpool and Merseyside* (1969).

Baptiste, Jean and Delambre, J., *Historie de l'Astronomie Moderne, II* (Paris, 1821), pp. 495–515.

Barocas, V., 'Jeremiah Horrocks (1619–1642)', *Journal of the British Astronomical Association*, 79 (1968–9), 223–6.

Bendall, S., Brooke, C., and Collinson, P., *A History of Emmanuel College, Cambridge* (Woodbridge: Boydell Press, 1999), chapter 3, 'Emmanuel and Cambridge: the early seventeenth-century hayday'.

Brahe, Tycho, *Tychonis Brahe Dani Opera Omnia*, ed., J. L. E. Dreyer (Copenhagen, 1913–29).

Britten, F. J., *Old Clocks and Watches and Their Makers*, sixth edition (London, 1932).

Bush Jr., Sargent and Rasmussen, Carl J., *The Library of Emmanuel College, Cambridge*, 1584–1637 (Cambridge, 1986).

Camden, William, *Brittania* (English edition, 1607).

Chandler, George, *William Roscoe of Liverpool* (London, 1953).

Chapman, Allan, *Three North Country Astronomers* (Manchester, 1982) (Chapman I).

——, 'Jeremiah Horrocks, the Transit of Venus, and the New Astronomy in Early 17th-Century England', *Quarterly Journal of the Royal Astronomical Society*, 31 (1990), 333–57 (Chapman II).

——, 'Jeremiah Shakerley (1626-1655)', *Transactions of the Historic Society of Lancashire and Cheshire*, vol. 135 (1985) (Chapman III).

Copernicus, Nicholas, *De Revolutionibus Orbium Coelestium* (Poland, 1542).

Crossley, James (ed.), 'Diary and Correspondence of Dr Worthington', *Chetham Society*, vol. XIII (1847).

Davis, Betty M., 'The Astronomical Work of Jeremiah Horrocks', University of London M.Sc. thesis 1967.

Dictionary of National Biography (repr. London: Oxford University Press, 1949–50), 9, 1267–9. (*DNB*)

Fernie, J.D. , 'The extraordinary and short lived career of Jeremiah Horrocks', AM SCI 84 (2): 114-117 MAR-APR 1996.

Flamsteed, John, *The Correspondence of John Flamsteed, the first Astronomer Royal*, ed. Forbes, Murdin and Wilmoth, 3 vols (Cambridge, 2001).

Forbes E. G., *The Gresham College Lectures of John Flamsteed* (London, 1975).

Fuller, Thomas, *The History of the University of Cambridge from the Conquest to the year 1634*, ed. Prickett and Wright (Cambridge, 1840).

Galilei, Galileo, *Dialogo sopra i due massimi sistemi del mondo, ptolemaico e copernicano* (Dialogue Concerning the Two Chief World Systems, Ptolemaic & Copernican, published 1632).

Gaythorpe, Sidney B., 'Jeremiah Horrocks and his "New Theory of the Moon"', *Journal of the British Astronomical Association*, 67 (1957), 134–44 (Gaythorpe I).

——, 'On Horrocks's Treatment of the Eviction and the Equation of the Centre . . .', *Monthly Notices of the Royal Astronomical Society*, 85 (1925), 858–65 (Gaythorpe II).

—, 'Horrocks's Observations of the Transit of Venus, 1639 November 24 (O.S.)', *Journal of the British Astronomical Association*, 47 (1936–7), 60–8; 64 (1953–4), 309–15 (Gaythorpe III).

—, 'Jeremiah Horrocks: Date of Birth, Parentage and Family Associations', *Transactions of the Historical Society of Lancashire and Cheshire*, 106 (1954), 23–33. (Gaythorpe IV).

—, 'Horrocks's Observations and Contemporary Ephemerides', *Journal of the British Astronomical Association*, 47 (1937), 156–7 (Gaythorpe V).

—, 'Horrocks's Observations of the Transit of Venus 1639 November 24 (O.S.)', *Journal of the British Astronomical Association*, 47 (1936), 60–8; 64 (1954), 309–15 (Gaythorpe VI).

Grant, Robert, *History of Physical Astronomy from the Earliest Ages to the Middle of the Nineteenth Century* (London, 1852), pp 420–28, 545.

Hall, Lawrence, 'The Ancient Chapel of Toxteth Park and Toxteth School', *Transactions of the Historical Society of Lancashire and Cheshire*, 87 (1935), 23–57 (Hall I).

—, 'The birth-place of Jeremiah Horrocks in Toxteth Park', *Transactions of the Historical Society of Lancashire and Cheshire*, 88 (1936), 249–55 (Hall II).

Halley, Robert, *Lancashire, its Puritanism and Nonconformity* (London, 1869).

Horrocks, Jeremiah, *Venus in Sole Visa* (Danzig 1662). The text used is the English translation by A. B. Whatton (see Whatton) with page numbers referenced from the same (Horrocks, *Venus*).

—, 'Astronomical Exercises', RGO Flamsteed's Papers 1/68A (Horrocks, 'Astronomical Exercises').

—, 'Philosophical Exercises', RGO Flamsteed's Papers 1/68B (Horrocks, 'Philosophical Exercises').

Kepler, Johannes, *Mysterium Cosmographicum* (1596).

Lansberg, P., *Tabulae perpetuae*, Trinity College Library, Cambridge (1635).

Newton, Isaac, *(The) Correspondence of Isaac Newton*, ed. H. W. Turnbull, J. F. Scott, A. R. Hall, L. Tilling, vols I–VII (Cambridge, 1959–77) (Newton, *Correspondence*).

—, *Philosophiae Naturalis Principia Mathematica* (1st edition London 1686) (Newton, *Principia*).

Opera Posthuma, see Wallis.

Plummer, H.C., 'Jeremiah Horrocks and his Opera Posthuma', *Notes and Records of the Royal Society*, 3 (1940-1), 39-52.

Public Records Office, London, Original Works. Horrocks's surviving manuscripts kept with Flamsteed's papers, vols. LXVIII and LXXVI on microfilm.

Rigaud, Stephen P. and Stephen J., *Correspondence of Scientific Men of the Seventeenth Century* (Oxford, 1841). Letters of Wallis, Flamsteed and Newton on Horrocks, vol. 2, 108–20, 338.

Royal Greenwich Observatory, RGO Archives, Flamsteed Papers RGO 36/22 vols 67–9 and RGO 36/3 vols 7–14.

Scriba C. J., 'The autobiography of John Wallis FRS', *Notes and Records of the Royal Society*, xxv (1970).

Spaulding W. F., 'A Country Curate', *Quarterly Journal of the Royal Astronomical Society*, 12 (1971), 179–82.

Twemlow J. A. (ed.), *Liverpool Town Books 2 (1571–1603)* (Liverpool, 1938).

Victoria County History, *Lancashire* (Cambridge, 2001), vol. 3 (London, 1908).

Wallis, John (ed.), *Opera Posthuma of Jeremiah Horrocks* (*Opuscula astronomica*) (London, 1672–3, 1678).

Watson, E. C., 'An Interesting Tercentenary', *Publications of the Astronomical Society of the Pacific*, 51 (1939), 305–14.

Webster, Charles, 'The Towneley Group', *Transactions of the Historical Society of Lancashire and Cheshire*, vol. 18 (1966).

Whatton A. B., *Memoir of the Life and Labours of the Reverend Jeremiah Horrocks* (London,1859), including a translation of Horrocks's *Venus in Sole Visa.*

INDEX

The letter 'p' before the page number denotes a picture.